mainly produces various species of carp and one species of tilapia (Nile tilapia) (see Table I.1). Aquaculture production of marine species is distinctly classified as marine fish farming.

Fish species	Live weight tonnage (thousands of tons)
Grass carp (*Ctenopharyngodon idella*)	5704.0
Silver carp (*Hypophthalmichthys molitrix)*	4788.5
Nile tilapia (*Oreochromis niloticus*)	4525.4
Common carp (*Cyprinus carpio)*	4189.5
Atlantic salmon (*Salmo salar*)	2435.9
Rainbow trout (*Oncorhynchus mykiss*)	848.1
Marine fish	767.5

Table I.1. *World production in 2018 of the main fish species (according to FAO (2020))*

The rise of fish farming has been based on the breeding of species such as tilapia, carp and certain salmonids such as trout or salmon. Alongside traditional fish farming, a new concept called the "blue revolution" based on the genetic transformation of these species has also been developed over the past 30 years. Genetic engineering, a tool for transgenesis, can also be applied to other aquatic resources such as microalgae for the purpose of producing molecules of therapeutic interest, for example for antivirals (Fleurence 2021a).

This biotechnological approach involving marine or freshwater organisms is referred to as "blue biotechnology". However, it is not limited to genetic engineering applied to transgenesis. Indeed, the use of enzymatic or microbiological engineering to transform the nutritional or organoleptic properties of a marine food resource are also considered biotechnological approaches and are included in the concept of blue biotechnologies. The latter are notably applied to macroalgae or fish to improve the nutritional or taste qualities of products offered for human consumption.

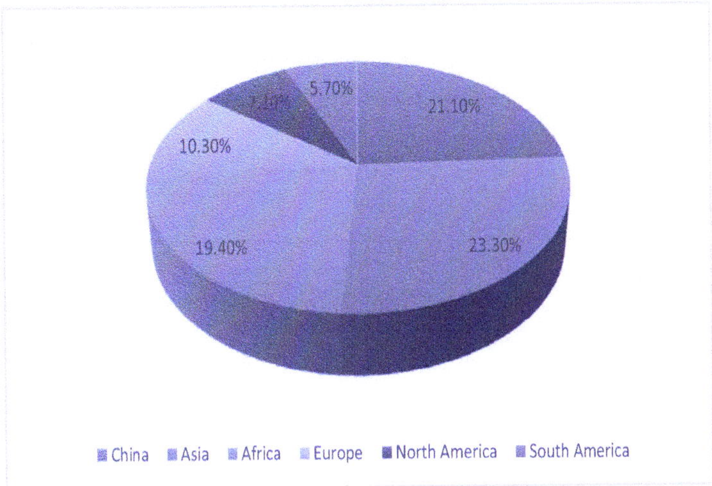

Figure I.1. *Distribution of animal protein intake through fish consumption in different regions of the world (according to Tacon (2003)). For a color version of this figure, see www.iste.co.uk/fleurence/aquatic.zip*

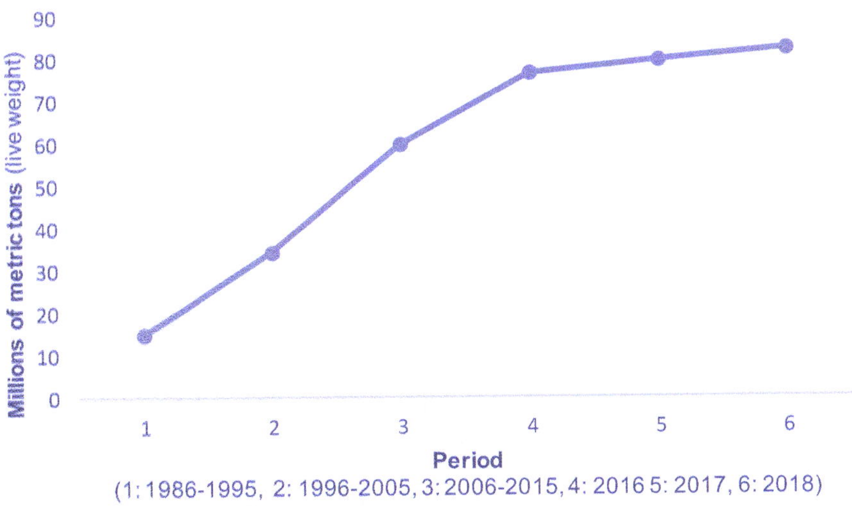

Figure I.2. *Trends over the last four decades for world aquaculture production of fish, including shellfish (according to FAO (2020))*

Fish production by aquaculture, also known as fish farming, today accounts for 52% of the volumes destined for human consumption (FAO 2020). The farming of freshwater species, known as continental fish farming,

Introduction

Aquatic resources play an important role in the overall dietary intake of many populations. This is particularly true in China, where fish accounted for just over 21% of animal protein intake in the population diet in the early 2000s (see Figure I.1) (Tacon 2003). On the Asian continent as a whole, fish consumption accounted for 23% of dietary animal protein intake (see Figure I.1).

Since the early 1960s, the annual growth rate in fish consumption has been 3.1%, whereas the annual increase in the human population has been half that (1.6%) (FAO 2020). This annual increase in fish consumption is significantly higher than that of other sources of animal food proteins, such as meat, milk or their derivatives (+2.1%).

Today, this trend translates into a per capita consumption of 30–50 kg of fish per year in Asia (FAO 2020). Maritime Europe, including countries such as France, Spain and Scandinavia, shows a fish consumption rate comparable to those reported for the Asian continent (FAO 2020). In contrast, some countries such as Greenland, Norway and Iceland have fish consumption levels well in excess of 50 kg per capita per year (FAO 2020).

The inclusion of fish as a source of animal protein in the human diet has therefore helped to meet the demographic challenges facing the world's population. This food security challenge has been met by increasing fish catches, and also by the unprecedented development of aquaculture over the last four decades (see Figure I.2).

Marine biotechnologies or blue biotechnologies are not limited to transgenesis applied to marine organisms, whether of animal or plant origin.

The application of technology to living organisms includes all the processes likely to change the characteristics of a biological matrix by modifying behavior, nutritional value or organoleptic quality.

In this context, this book also deals with enzymatic or fermentative transformation processes that can be applied to an animal matrix (fish fillets) or plant matrix (seaweed), in order to improve the nutritional or organoleptic added value of the final product. Particular emphasis is placed on the contribution of enzymatic engineering to the elimination of anti-nutritional compounds in seaweed and the development of new flavors in the case of fish fillets and seaweed.

In addition to the biotechnological aspects, the book also reviews the regulatory obstacles that may apply, depending on the country, to products derived from blue biotechnology and more particularly to genetically modified organisms.

More generally, the aim of this book is to take stock of aquaculture production methods incorporating transgenesis as a biotechnological tool and of biotechnological processing methods applied to marine resources, whatever their mode of production (fishing, traditional aquaculture, "blue revolution" aquaculture).

As such, it is particularly aimed at teacher-researchers, teachers and engineers working in the fields of aquaculture or seafood product valorization. It is also of interest to students and engineering students training in these fields.

November 2023

Preface

Since the end of World War II, humankind has been trying to cope with a growing population. The animal protein production system based on terrestrial animal husbandry soon proved insufficient to meet food needs generated by the demographic growth of many Asian countries such as China and India. This food security challenge has been largely met by the development of aquaculture, and in particular fish farming. The breeding of freshwater fish, such as tilapia and trout, is at the root of the worldwide boom in inland fish farming. In marine fish farming, Atlantic salmon, a diadromous species with anadromous behavior, has rapidly established itself as a development and production model for this type of aquaculture. Genetic engineering, a biotechnological process subject to much controversy, has enabled the production of transgenic fish from species commonly used in fish farming.

This book takes stock of the biological characteristics of the fish concerned, their traditional farming methods and the contribution of biotechnologies to the development of a new aquaculture described by its initiators as the "blue revolution".

However, the development of transgenesis is not limited to marine organisms of animal origin. More recently, genetically modified microalgae have been used to create humanized antibodies and recombinant vaccines. This biotechnological approach, which has been little developed at the industrial level, is nonetheless an interesting example of the application of transgenesis. As such, it is discussed briefly in this book.

Contents

First published 2024 in Great Britain and the United States by ISTE Ltd and John Wiley & Sons, Inc.

ISTE Ltd
27-37 St George's Road
London SW19 4EU
UK

www.iste.co.uk

John Wiley & Sons, Inc.
111 River Street
Hoboken, NJ 07030
USA

www.wiley.com

Library of Congress Control Number: 2023949514

British Library Cataloguing-in-Publication Data
A CIP record for this book is available from the British Library
ISBN 978-1-78630-969-3

Aquatic Biotechnologies

From Genetic Engineering to Enzymatic or Fermentation Engineering

Joël Fleurence

WILEY

Aquatic Biotechnologies

1

The Biological Characteristics of Organisms Involved in Blue Biotechnologies

1.1. Fish

1.1.1. *Tilapia*

The genus *Tilapia* includes some 40 species from East Africa. In the vernacular, the term "tilapia" is used to designate species belonging to the genus *Tilapia* and also to the genera *Oreochromis* and *Sarotherodon*. Two species are mainly exploited for human consumption: Nile tilapia (*Oreochromis niloticus*) (see Figure 1.1) and Mozambique tilapia (*Oreochromis mossambicus*). Tilapia are freshwater fish with an omnivorous diet, but with a strong herbivorous tendency.

Nile tilapia is the most consumed species and contributes to the diet of many populations on the African and Asian continents. This species, raised in Ancient Egypt, was able, after introduction, to adapt to different ecosystems far from its continent of origin (Japan, Thailand, Vietnam, Brazil, China). It is a tropical fish that lives in shallow waters and has a strong aptitude for colonizing ponds and streams. However, the Nile tilapia is very sensitive to water temperature. Temperatures below 11°C and above 42°C prove to be lethal for this organism. Water depth is also a parameter that plays a significant role in mortality (El-Sayed et al. 1996). A depth of 200–300 cm reduces mortality observed for fish living at a depth of 50 cm by approximately 50% (see Figure 1.2).

This is an important criterion, particularly for breeding this species. In addition to habitat depth, temperature has a major influence on growth (see Figure 1.3). It also plays a role in the reproductive process, since spawning, after sexual maturation, is only possible when water temperature reaches 24°C.

Figure 1.1. Oreochromis niloticus *or Nile tilapia (photo credit © Tørrissen B.C. 2012, Wikimedia/CC BY-SA 3.0)*

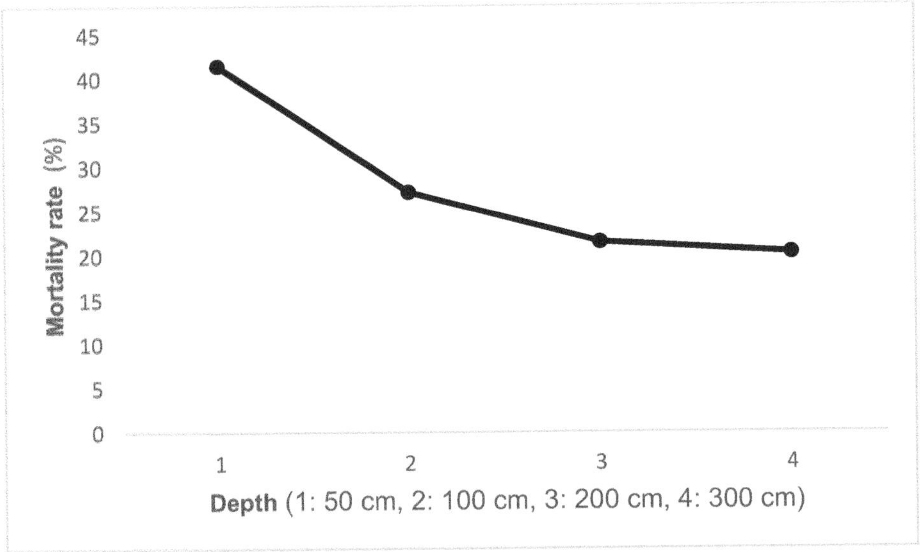

Figure 1.2. *Impact of pond depth on the survival of Nile tilapia (*Oreochromis niloticus*) (average over 10 months of rearing) (from El-Sayed et al. (1996))*

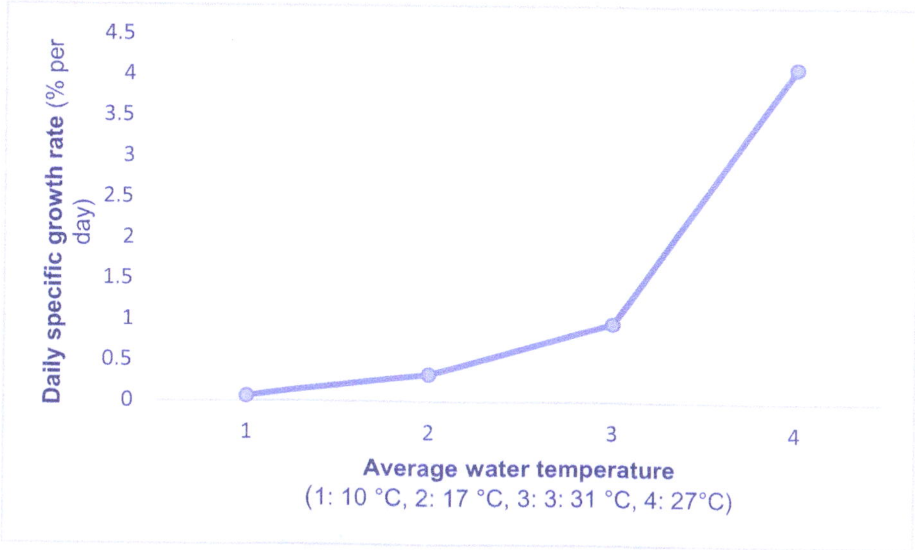

Figure 1.3. *Impact of pond temperature on the specific growth rate of Nile tilapia (*Oreochromis niloticus*) at a depth of 3 meters (from El-Sayed et al. (1996))*

1.1.2. *Trout*

"Trout" is the common name for several species of fish belonging to the Salmonidae group. Two distinct taxa or genera, the genus *Oncorhynchus* and the genus *Salmo*, are listed principally. A third, the genus *Salvelinus*, including the species *Salvelinus fontinalis*, known as speckled or blue trout, is included in the common trout group. The common trout found in whitewater rivers generally belong to the *Salmo trutta* species (see Figure 1.4). In contrast, rainbow trout, which is used worldwide in aquaculture, is listed under the scientific name of *Oncorhynchus mykiss*. Other trout species, or even subspecies, are also harvested or exploited in aquaculture but are more limited than the previous two species. They live in a wide variety of environments, including whitewater rivers, lakes and the marine environment (see Table 1.1). Like salmon, some trout species are anadromous. This is particularly true of rainbow trout (*Oncorhynchus mykiss*), which leave the marine environment to reproduce in freshwater rivers. Trout are carnivorous fish, feeding on aquatic and aerial insects and various invertebrates. Examples of cannibalism by trout towards their fry have also been reported for some species and under some conditions

(Lestage 1937). This is particularly true of *Salmo trutta* and *Salvelinus alpinus* (Arctic trout) (Klemetsen et al. 2003).

Figure 1.4. Salmo trutta fario *or common trout (photo credit © Engbretson E. 2006, Wikimedia/public domain)*

Trout species or subspecies	Common name	Living environment
Oncorhynchus mykiss	Rainbow trout or salmon	Well-oxygenated seas and rivers
Oncorhynchus mykiss aguabonita	Golden trout	High-altitude lakes and rivers (> 2,100 m)
Oncorhynchus clarkii	Cutthroat trout	Seas (Alaska) and rivers (Northern California)
Salmo trutta trutta	Sea trout	Seas (Atlantic and Mediterranean)
Salmo trutta fario	Brown trout	Mountain streams, rivers, brooks
Salmo trutta lacustris	Silver trout	Lakes
Salmo trutta ferox	Ferox trout	Lakes (Iceland, Scotland)
Salmo trutta labrax	Black Sea salmon	Sea (Black, Azov, Caspian and Aral seas)

Table 1.1. *Examples of different trout species and their habitats (based on FAO (2020) and Wikipedia)*

1.1.3. *Atlantic salmon*

The term "salmon" is applied to various species belonging to the Salmonidae group. There are three main genera (*Oncorhynchus*, *Hucho*, *Salmo*), with varying numbers of species (see Table 1.2). The genus *Salmo* is represented by one species, *Salmo salar* (see Figure 1.5), known by its common name of Atlantic salmon. This species occurs naturally in the northern Atlantic zone and along its watershed. However, the significant development of aquaculture activity around this species has led to its presence in fish farms in the Southern Hemisphere, such as in Chile, which is one of the three main producers of Atlantic salmon, along with Norway and the United Kingdom. The genus *Hucho* comprises two species found mainly in the rivers of eastern Europe and the Far East (Siberia). Finally, the genus *Oncorhynchus* comprises eight species found in the northern part of the Pacific Ocean and its watershed. Most salmon species are anadromous, similar to some trout species (see section 1.2).

Figure 1.5. Salmo salar *or Atlantic salmon*
(photo credit © Fjeld H.P. 2006, Wikimedia/CC BY-SA 2.5)

Some species belonging to the genus *Oncorhynchus* may be known by more than one common name. This is particularly true of *Oncorhynchus keta*, which is known by four different common names (see Table 1.2).

Salmon is a carnivorous fish that feeds on small aquatic insects and mayflies in freshwater. In the marine environment, their diet is based on the predation of various crustaceans, including shrimps rich in astaxanthin, the pigment that gives salmon its red flesh. They also feed on fry or small fish of different species. Independently of this aspect, cannibalism is also known to exist in the *Salmo salar* species (Klemetsen et al. 2003).

Salmon species	Common names	Natural geographical area
Salmo salar	Atlantic salmon	North Atlantic and watershed
Oncorhynchus kisutch	Silver salmon Coho salmon	North Pacific and watershed
Oncorhynchus keta	Dog salmon Pacific salmon Keta salmon Chum salmon	North Pacific and watershed
Oncorhynchus tshawytscha	Chinook salmon King salmon	North Pacific and watershed
Oncorhynchus masou masou	Japanese salmon Masu salmon	Far East
Oncorhynchus nerka	Nerka salmon Red salmon	North Pacific and watershed
Oncorhynchus gorbuscha	Pink humpback salmon	North Pacific and watershed
Hucho hucho	Danube salmon Huchen	Eastern and northern Europe

Table 1.2. *Examples of species grouped under the generic name of salmon and natural geographical areas (from Wikipedia)*

1.1.4. *Carp*

As before, the common name carp encompasses a number of different fish species. These belong to the Cyprinidae group. The common carp is identified by its Latin name, *Cyprinus carpio*. Other species with multiple common names are also listed (see Table 1.3). In particular, the name "Asian carp" covers several different species, which have in common that they live in the geographical area of the Asian continent or Eastern Siberia. Carp are freshwater fish that live in the middle and lower reaches of rivers. In addition to this habitat, the common carp is also found in the wild in flooded areas, lakes and reservoir waters. They are omnivorous fish, feeding mainly on plant debris and plants (algae), as well as on mollusks and invertebrate larvae (insects, worms and crustaceans).

Carp species or subspecies	Common names	Natural geographical area
Cyprinus carpio	Common carp	Europe
Cyprinus carpio carpio	Leather carp Japanese carp Koi carp Mirror carp	Asia
Hypophthalmichthys molitrix	Silver carp Silver love carp Asian carp Chinese carp	China, Eastern Siberia
Ctenopharyngodon idella	Grass carp Herbivorous carp Amour carp Asian carp	Amur River (China–Siberia)
Mylopharyngodon piceus	Black carp Asian carp	China, Vietnam, Siberia
Hypophthalmichthys nobilis	Bighead carp Marbled carp Asian carp	Asia, North America

Table 1.3. *Some examples of species grouped under the common name of carp (from Wikipedia)*

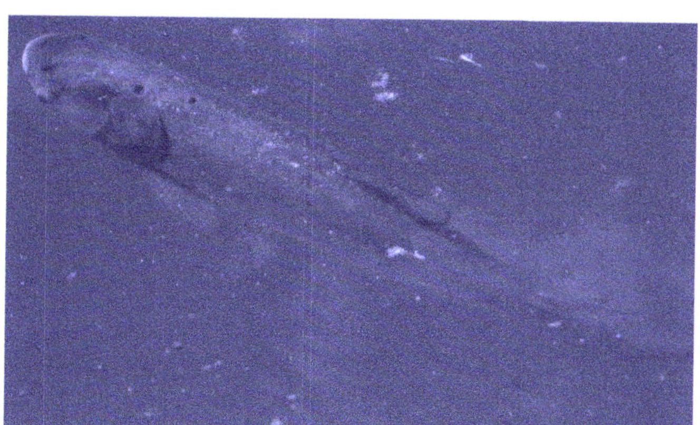

Figure 1.6. *Hypophthalmichthys molitrix or silver carp (photo credit © Strauss T. 2006, Wikimedia/CC BY-SA 3.0)*

1.2. Algae

Algae are a marine resource often used in blue biotechnologies. They are eukaryotic photosynthetic organisms found mainly in aquatic environments and also in other particularly humid environments.

A distinction is made between multi-cellular algae or macroalgae and unicellular algae, known as microalgae.

Macroalgae are often transformed by biotechnological processes (enzymatic hydrolysis, fermentation, etc.) for food applications. Microalgae are subjected to other biotechnological processes, such as genetic transformation to produce new therapeutic molecules (Fleurence 2021a).

1.2.1. *Microalgae*

Microalgae are unicellular, eukaryotic organisms that can associate to form colonies. Botanically, they are categorized into three phyla: Chlorophytes (green algae), Rhodophytes (red algae) and Chromophytes (golden brown algae) (Fleurence 2021a) (see Figures 1.7–1.9).

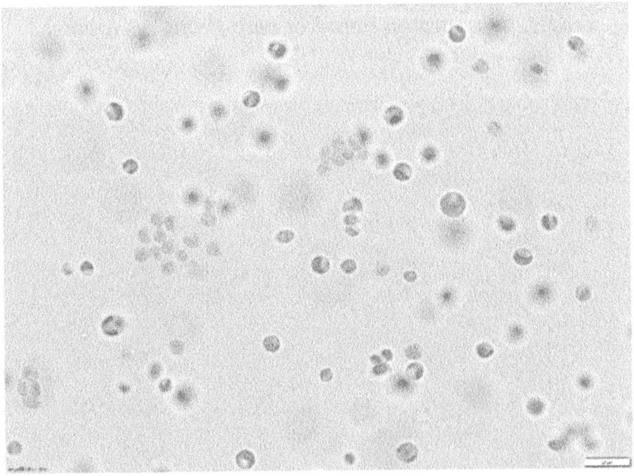

Figure 1.7. *Example of a Chlorophyte:* Chlorella vulgaris *or* Chlorella *(scale: 20 µm) (photo credit © Isomer, Petit A. 2022)*

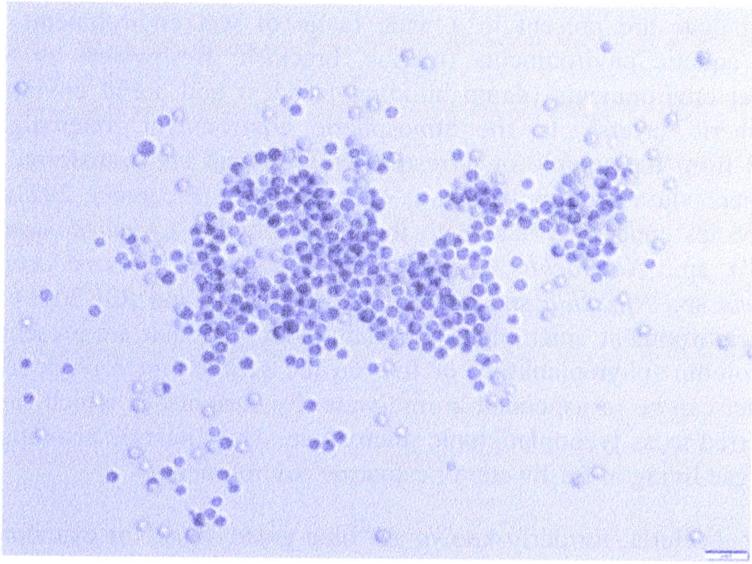

Figure 1.8. *Example of a Rhodophyte:* Porphyridium cruentum *or* Porphyridium *(scale: 20 μm) (photo credit © Isomer, Petit. A. 2021)*

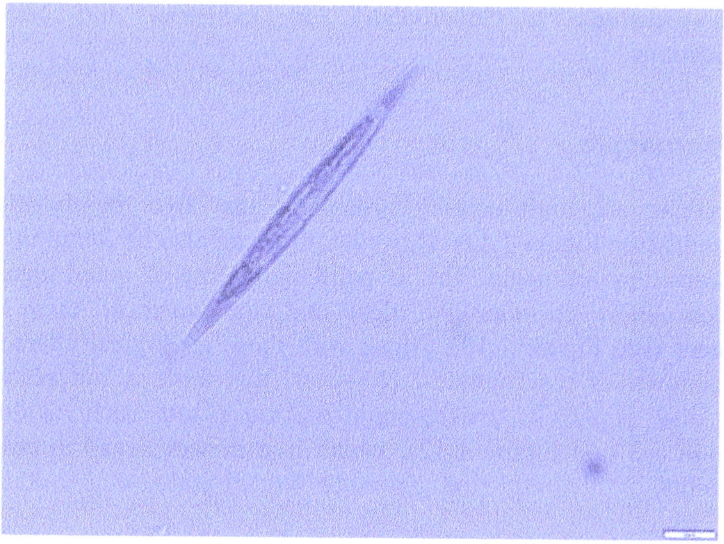

Figure 1.9. *Example of a Chromophyte:* Haslea ostrearia *(diatom) (scale: 10 μm) (photo credit © Isomer, Petit. A. 2019)*

Microalgae are present in a wide range of wet environments. These include aquatic environments (marine, brackish, freshwater) as well as terrestrial environments (damp building facades) and aerial environments (atmospheric layers). In the atmospheric environment, microalgae are released from the aquatic or terrestrial environment and transferred to the atmosphere via a process known as "aerolization" (Fleurence 2021a). The main species concerned belong to the Chlorophytes (*Chlamydomonas* sp., *Chlorella* sp., *Nannochloris* sp.) or Chromophytes (*Heterococcus* sp., *Vaucheria* sp., *Nitzschia* sp., *Navicula* sp.) (Sharma and Rai 2011). In the marine environment, microalgae and diatoms in particular are present in the water column (phytoplankton) or live on the seabed (microphytobenthos). The latter can be resuspended in mid-water by currents, in which case they are referred to as tychoplanktonic microalgae. This lifestyle also applies to microalgae living in freshwater or estuarine environments.

Cyanobacteria, formerly known as "blue-green algae" or cyanophyceae, are not considered microalgae, since they are prokaryotic organisms. From an evolutionary point of view, they belong to the eubacteria, not the eukaryotes, as in the case of microalgae. Despite this, some species, such as *Arthrospira platensis* (*Spirulina platensis*), better known as Spirulina, are still presented as "microalgae" in industrial and commercial communications.

1.2.2. *Macroalgae*

Macroalgae are multi-cellular organisms that live mainly in marine environments (see Figure 1.11). However, they can also be found in brackish or freshwater environments. This is particularly true of green algae of the genus *Cladophora*, cosmopolitan algae that can live in all three types of environment (see Figure 1.10). Algae with these ecological characteristics are also known as macrophytes. However, this term is not restricted to algae. It also applies to macroscopic aquatic plants such as duckweed (see section 1.3 and Figure 1.12), which live in wet zones in continental environments.

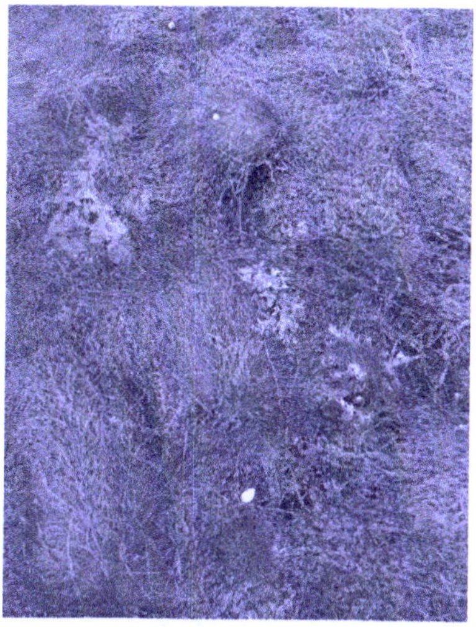

Figure 1.10. *Example of a cosmopolitan green alga (marine, brackish, freshwater)* Cladophora *sp. (photo credit © Fleurence J. 2022)*

Figure 1.11. *Examples of Rhodophytes (*Palmaria palmata*), Chlorophytes (*Ulva *sp.) and Chromophytes (*Fucus *sp.) on the foreshore (photo credit © Fleurence J. 2022)*

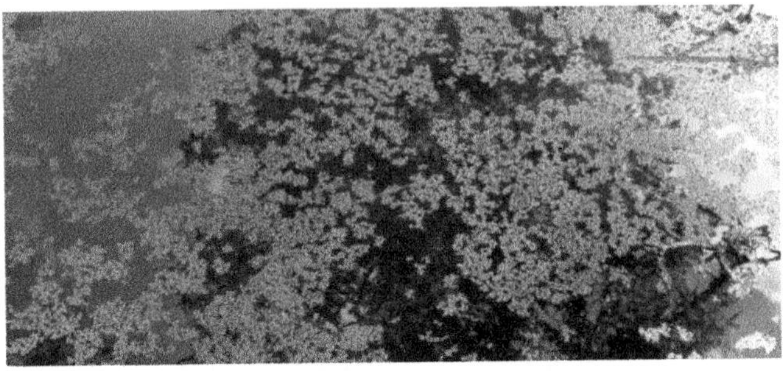

Figure 1.12. *Example of a freshwater macrophyte, duckweed* (Lemna minor) *(photo credit © Fleurence J. 2022)*

Just like microalgae, multi-cellular algae belong to three phyla: Chlorophytes, Rhodophytes and Chromophytes (see Figure 1.11).

1.3. Other aquatic plants

Some macrophyte species, found mainly in freshwater, are the subject of recent interest in human nutrition (Appenroth et al. 2018). These include duckweed in particular. The name "duckweed" is a common term applied to certain aquatic plants whose morphology is similar to that observed in duckweeds. This name covers three main genera: *Lemna*, *Spirodela* and *Wolffia*.

Duckweeds are aquatic organisms with a worldwide distribution. Thanks to their mode of reproduction, which is mainly based on vegetative multiplication, these organisms can proliferate to excess depending on favorable environmental conditions (temperature, light). In this case, the duckweed forms a carpet quickly covering the surface of the watercourse and more particularly the surfaces of stagnant water. The most widespread and cosmopolitan species is called *Lemna minor* or small duckweed. It is characterized by the presence of oval leaves, of just under a centimeter in size. It is a phanerogam whose flowers have a size of the order of a millimeter. As a phanerogam, duckweed is distinguished from algae, which lack floral organs.

Species of the genus *Wolffia*, and *W. globosa* in particular, are currently cultivated and processed for human and animal consumption.

2

Production Methods

2.1. Fish farming

Fish farming is an aquaculture specialty. It concerns the rearing of fish in freshwater, brackish or marine environments. Fish may be reared in basins or in a natural environment (ponds, sea). In the open sea, fish are bred in fish cages.

2.1.1. *Traditional fish farming*

Freshwater fish farming was first developed in China and India around 1,500 BCE. A Chinese treatise on fish farming entitled "The Rearing of Tao Zhu Gong Fish", or Fan Li's treatise, was written in the 5th century BCE. This work compiles a series of interesting experiments on the rearing of carp in ponds. From a historical point of view, freshwater fish farming represents the oldest traditional fish farming activity.

It is carried out in rivers, ponds or lakes. As such, it is called inland fish farming. Fish farming in marine environments developed later, in the middle of the 20th century. It is known as marine fish farming and also represents a traditional fish farming activity.

Aquatic animal farming is dominated worldwide by fish production (54.3 million tons) (FAO 2020), which accounts for 60% of global aquaculture production (see Figure 2.1).

Inland fish production has increased almost fivefold since 1990 (see Figure 2.2) (FAO 2020).

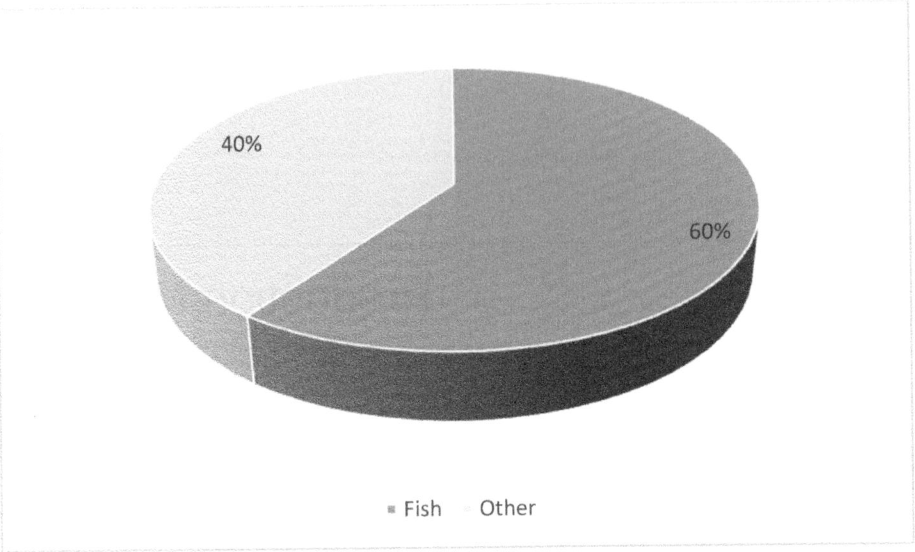

Figure 2.1. *Global aquaculture production split between fish and other animals (crustaceans, mollusks, echinoderms) (based on FAO (2020)). For a color version of this figure, see www.iste.co.uk/fleurence/aquatic.zip*

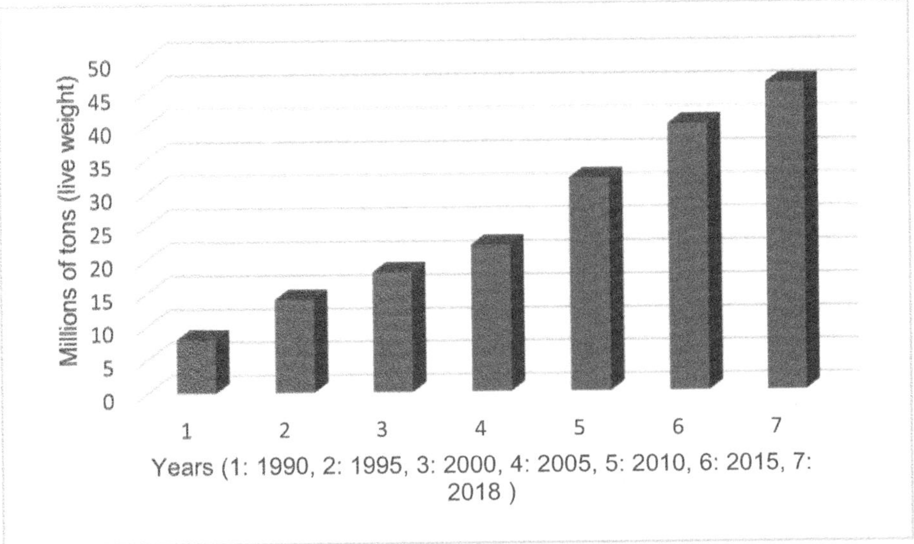

Figure 2.2. *Worldwide trends in inland fish farming between 1990 and 2018 (according to FAO (2020))*

In 2018, fish production accounted for 62.5% of global fish production. However, this growth slowed between 2017 and 2018 due to the decline in aquaculture activity in China, the world's leading producer. China is the world's leading producer of Cyprinidae (carp), including grass carp (*Ctenopharyngodon idella*), silver carp (*Hypophthalmichthys molitrix*), bighead carp (*Hypophthalmichthys nobilis*) and common carp (*Cyprinus carpio)* (FAO 2022a). Carp are reared in a confined environment, usually ponds of varying types (see Table 2.1). These ponds can be of natural origin or shaped by human activity. The water quality and nutritional richness required for fish rearing are highly dependent on the type of pond. For carp farming, ponds are stocked with small crustaceans belonging to the genera *Moina* or *Daphnia*. In the case of omnivorous carp, the juveniles are also fed a supplement of soya, cereals and sometimes meat. This addition of prey or supplementary food is known as pond fertilization. Oxygenation of the environment, a critical stage in this breeding technique, is achieved by abundant fresh water entering the pond and eliminating excess water through overflow. This release of partially used water into the surrounding environment is one of the major ecological drawbacks of this type of fish farming.

Pond types	Advantages/disadvantages
Spring pond	Spring water supply Water quality Source drying up
Groundwater pond	Groundwater recharge Water quality Groundwater depletion
Stormwater pond	Rainwater supply Water quality Drought drying
Stream-fed pond	Water supply by diverting streams and backwaters Earthworks requiring agricultural engineering work

Table 2.1. *Examples of pond types that can be used in traditional fish farming for rearing freshwater fish such as tilapia or carp*

Carp farming can also be combined with rice cultivation (Asia, Madagascar); this coupled activity is known as rizipisciculture (see

Figure 2.3). Fingerlings are brought in once the rice seedlings have taken root and the water is no longer muddy. At the end of the rice development cycle and eight days before harvest, the fish gather in the deepest areas of the pond. They can then be harvested for consumption, marketed or transferred to small storage ponds to continue breeding out of season for rice production. This production technique, although ancestral, is a traditional production method still used in many countries. Cyprinids can also be reared on land in ponds called viviers. On the other hand, in open environments such as rivers and streams, they are raised in floating net cages.

Pond rearing with or without rice cultivation is also traditionally used for rearing tipalia (*Oreochromis niloticus* or Nile tilapia). This species, called aquatic chicken, is currently the second most important fish species after carp. As with carp production, several types of pond are used (see Table 2.1). In addition to natural ponds, some fish ponds are artificially constructed. Artificial ponds are designed to meet particular specifications recommended by the FAO[1]. Ponds are square-shaped and 80–100 cm deep. They are fed by a canal, which should be at least 70 cm higher than the bottom of the pond. The bottom must itself be lower than the waterway used to drain the pond. The pond is protected by nets to prevent predation, particularly by birds (see Figure 2.3).

Figure 2.3. *Artificial rice–fish ponds for tilapia farming in Indonesia (photo credit © FAO aquaculture photo library, Stankus A. 2016)*

1 See: https://www.doc-developpement-durable.org/file/Elevages/aquaculture&peche&pisciculture/tilapia/tilapia-etang.pdf.

Apart from the fry stage, tilapia or carp can also be grown in open environments such as lakes or rivers. In this case, the animals concerned are reared in floating cages forming aquaculture parks.

Fish farming in natural environments, whether confined or not, poses the problem of the discharge of aquaculture wastewater into the environment and its eutrophying impact on endogenous plant and animal species. This type of fish farming is referred to as open-system fish farming. However, it can be improved by introducing mixed or multi-trophic cultures. This type of farming relies on recycling inorganic waste produced by fish (NO_3^-, NO_2^-, NH_4^+), plants and algae. In this sense, rice–fish farming can be considered as a more environmentally friendly form of production. Despite this example, fish farming in open systems, whether inland or marine, remains a non-negligible source of eutrophication of the environment. In contrast, semi-open or closed-system aquaculture based on the reprocessing of rearing water, although more costly, has less impact on the aquatic ecosystem.

More generally, inland fish farming can be based on four production models. These models correspond to types of production (extensive, semi-intensive, intensive) (see Table 2.2) and economic needs, as well as to customary or modern practices.

Production models	Production types	Equipment types	Species
Water system	Semi-intensive Intensive	Cages Enclosures Acadjas (African basket of branches)	Carp Tilapia
Earth system	Extensive Semi-intensive Intensive	Ponds Irrigated basins Raceways (masonry basins with water circulation)	Carp Tilapia Trout
Closed system	Intensive	Ponds with wastewater recycling system and recirculation of treated water into the production system	Carp Tilapia Trout
Multi-trophic system	Extensive Semi-intensive Intensive	Coupling of fish farming with cultivation of a plant species (rice–fish culture)	Carp Tilapia

Table 2.2. *Examples of inland fish farming production methods*

In Europe, inland fish farming is mainly based on salmonid farming, and trout farming in particular. Rainbow trout (*Oncorhynchus mykiss*) is the main species farmed. This is also the case worldwide, where its production accounts for 97% of the trout volume generated by aquaculture (see Figure 2.4) (Eumofa 2021). This species can be farmed in freshwater or marine environments. In Europe, it is mainly produced in fresh water for the portioned trout market (200–300 g). On the other hand, the production of larger trout relies mainly on sea farming in submerged cages. In the latter case, the production method applied is that of marine rather than inland fish farming.

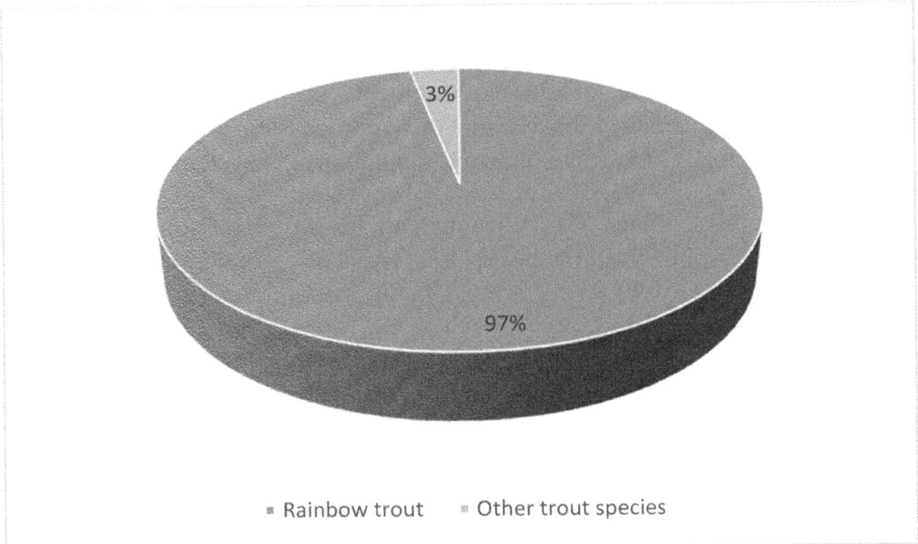

Figure 2.4. *Distribution of world rainbow trout production (*Oncorhynchus mykiss*) (according to Eumofa (2021)). For a color version of this figure, see www.iste.co.uk/fleurence/aquatic.zip*

Historically produced in Denmark, France, Germany, Italy and Spain, rainbow trout farming has recently expanded in Poland, where production capacity has increased by 15% in recent years (Eumofa 2017). Despite this, France, Denmark and Italy, with a production volume of 30,000 tons each, remain the main players in this aquaculture sector in Europe. Outside Europe, the leading producer of the species remains the Islamic Republic of Iran, which alone accounts for 22% of world production. Other major rainbow trout producers are Turkey, Norway, Chile and Peru.

Unlike tilapia, rainbow trout is farmed intensively in monocultures to make it economically viable.

Trout do not reproduce naturally in an aquaculture rearing system. The first step is to supply the system with juveniles. To do this, it is necessary to supply the eggs that will enable the juveniles to develop. The semen needed to produce these eggs is supplied by high-quality mature broodstock (three to four years old) (see Figure 2.5). Eggs are obtained by mixing male semen with oocytes previously collected from females. The fertilized eggs are then transferred and incubated in hatching trays or nesting boxes 4 meters long, 40 cm wide and 20 cm deep. The eggs are incubated in this structure until they reach the eyed stage (see Figure 2.5). At this stage, they can be sterilized and marketed to other producers who will continue the development process. Non-marketed eggs are kept in incubation until they hatch in the egg-laying chamber. Hatching results in the release of larvae with a yolk sac that acts as a food reserve. These larvae are referred to as alevins or young sac fry.

Egg production in the hatchery is one of the limiting stages in the provision of juvenile fish for farming. Indeed, eggs during this phase can be contaminated by fungal pathogens, and direct exposure to light proves lethal for embryos.

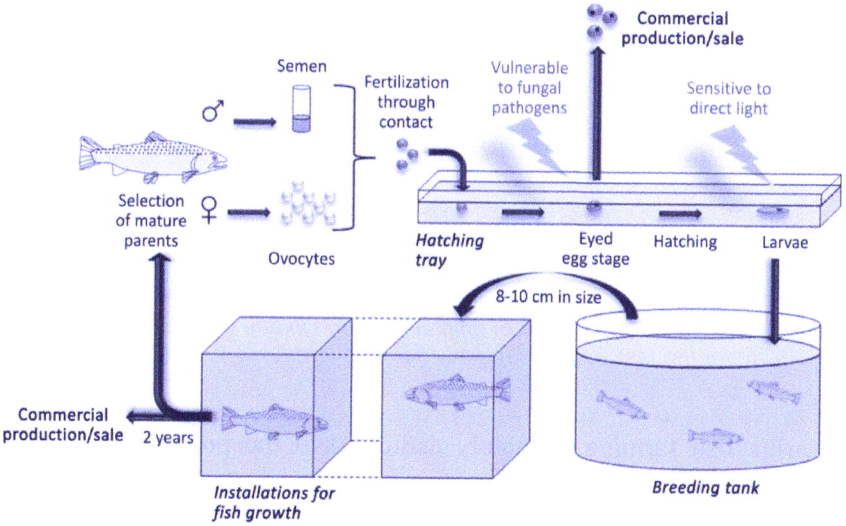

Figure 2.5. *Aquaculture production cycle for rainbow trout* (Oncorhynchus mykiss) *(Pouchus Y.F. 2022)*

Once the yolk sac has resorbed (two to four weeks), the fry have reached the "swimming" stage and will begin their foraging activity. At this stage, the fry are transferred from the rearing tanks. Once they have reached a size of 8–10 cm, the trout are moved to outdoor grow-out facilities, which may be concrete basins such as raceways or cages in ponds. The fish are raised in this way until they reach commercial size (30–40 cm). At this stage, they are marketed or released for recreational fishing. Some fish from the stock may also be selected as broodstock (see Figure 2.5).

The breeding of marine animals and more particularly of fish has been in constant progression for 30 years (+259%). However, over the past two decades, the gap between inland and marine aquaculture has steadily widened (see Figure 2.6). This aquaculture production, whether inland or marine, is essentially geared towards fish farming. Inland aquaculture is dominated by the production of carp and marine aquaculture by the farming of Atlantic salmon (FAO 2022a).

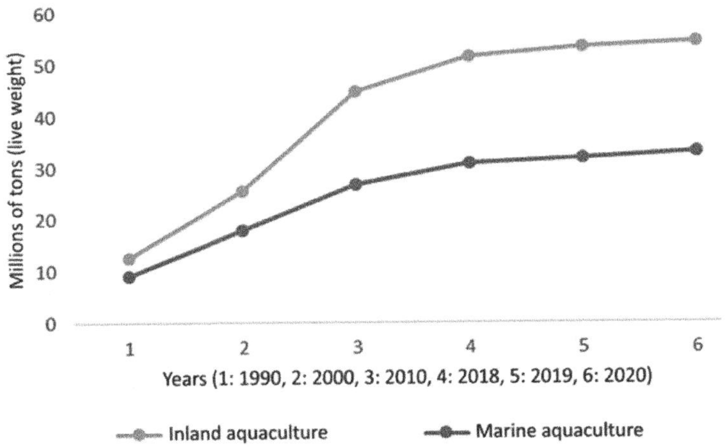

Figure 2.6. *World production of animals through inland or marine aquaculture over the last 30 years (mainly fish) (according to FAO (2022a))*

Marine fish farming is mainly dedicated to the production of Atlantic salmon (*Salmo salar*). The farming of salmon, and more specifically salmonids, is known as salmon farming.

The current world production exceeds 1,000,000 tons in live weight (FAO 2022a). Norway and Chile are the main producing countries (see Figure 2.7).

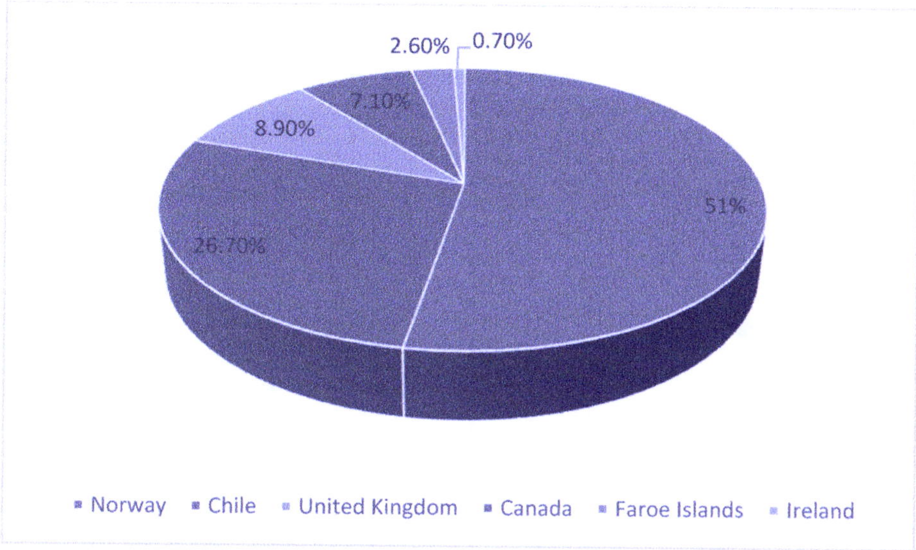

Figure 2.7. *Distribution of marine fish production of Atlantic salmon (*Salmo salar*) by main countries (according to Ifremer (2011)). For a color version of this figure, see www.iste.co.uk/fleurence/aquatic.zip*

Norway, the world's leading salmon producer, raises salmon in aquaculture sites that are well-suited to the purpose in terms of location (fjords, sparsely populated coastal strips) and climatic conditions.

Seawater rearing corresponds to the grow-out phase. Animals transferred to seawater are between 8 and 16 months old and weigh between 40 and 120 g. The earlier stages of development (eggs, fry, smolt and parr salmon) take place in fresh water in tanks or cages (see Figure 2.8). Growing animals are raised in marine environments in floating cages submerged in sheltered areas close to the coast (see Figure 2.9).

In Norway, salmon farming is subject to specific regulations and even standards. In particular, population density in the pens is subject to special regulations. In particular, the ratio of fish to water volume is limited to 2.5% salmon for 97.5% water[2].

2 See: https://toutsurlesaumon.fr/elevage-du-saumon-en-norvege/les-conditions-delevage-du-saumon/.

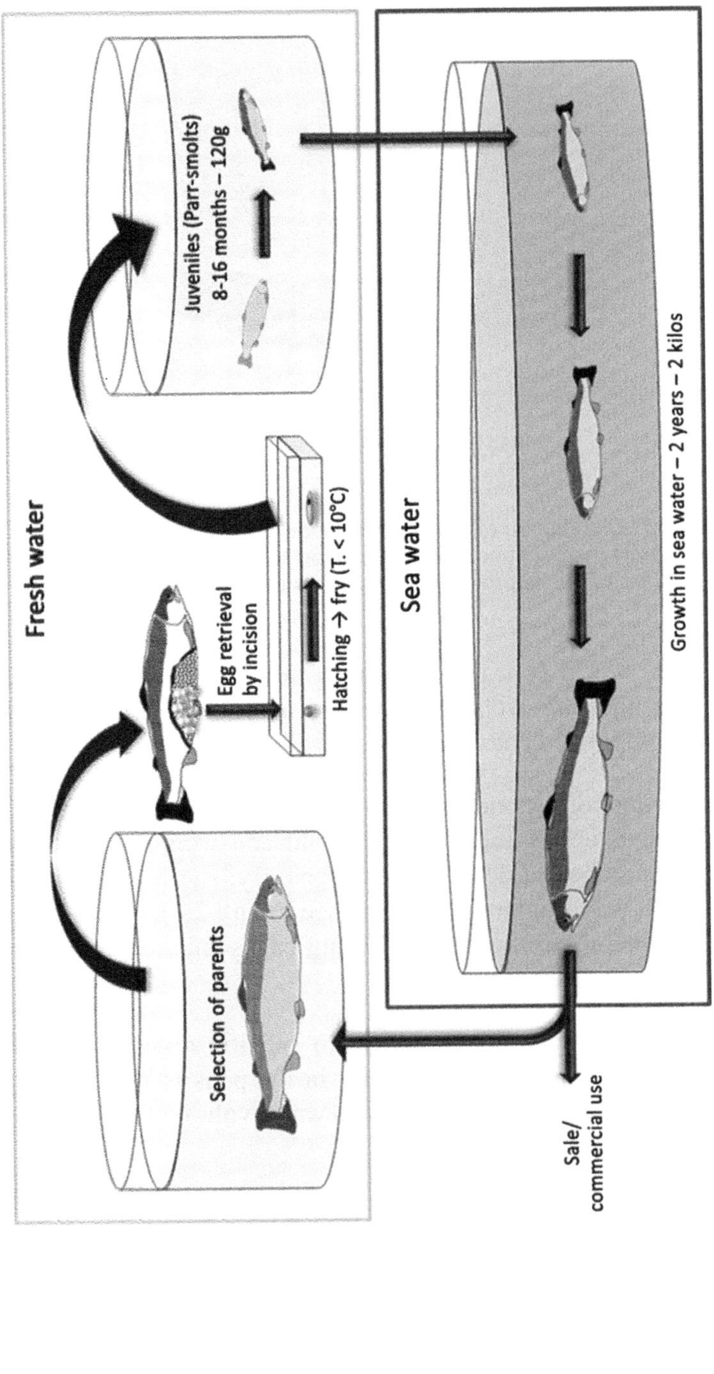

Figure 2.8. *Aquaculture production cycle for salmon* (Salmo salar) *(Pouchus Y.F. 2022)*

Figure 2.9. *Floating cages for salmon farming in Norway (photo credit © Stévant P. 2014)*

Since 2004, floating aquaculture sites have been subject to technical constraints designed to prevent salmon escaping into the natural environment. These constraints are governed by the NYTEK standard. Hygiene and infection control also form part of the specifications for salmon producers. A series of annual inspections, two of which are carried out by aquaculture veterinarians, are also part of salmon farming practices. A law on the welfare of fish produced has also been enacted. This covers not only farming but also transport and slaughter operations[3].

This last regulatory aspect of salmon farming in Norway is unique in world fish farming.

2.1.2. *Fish farming with transgenic fish or the concept of the blue revolution*

Fish farming with transgenic fish is a production method that is limited and can be described as experimental compared to traditional fish farming. It is based mainly on the genetically modified *Salmo salar* (Atlantic salmon) species. This transgenic salmon, produced by the American company AquaBounty Technologies, is known commercially as AquAdvantage

3 See: https://lovdata.no/dokument/NL/lov/2009-06-19-97.

salmon. The company that produces it describes itself as an expert in the rearing of transgenic fish in land-based aquaculture systems.

Historically speaking, the development of the transgenic salmon aquaculture industry has spanned some 30 years, from 1991 to the present day.

Production of genetically modified animals was initially carried out at sites in North America and Panama. The fertilized and genetically modified eggs were mainly produced in Canadian aquaculture facilities. These latter were located on Prince Edward Island in Rollo Bay at the mouth of the St. Lawrence River, close to the French archipelago of Saint-Pierre-et-Miquelon (see Figure 2.10).

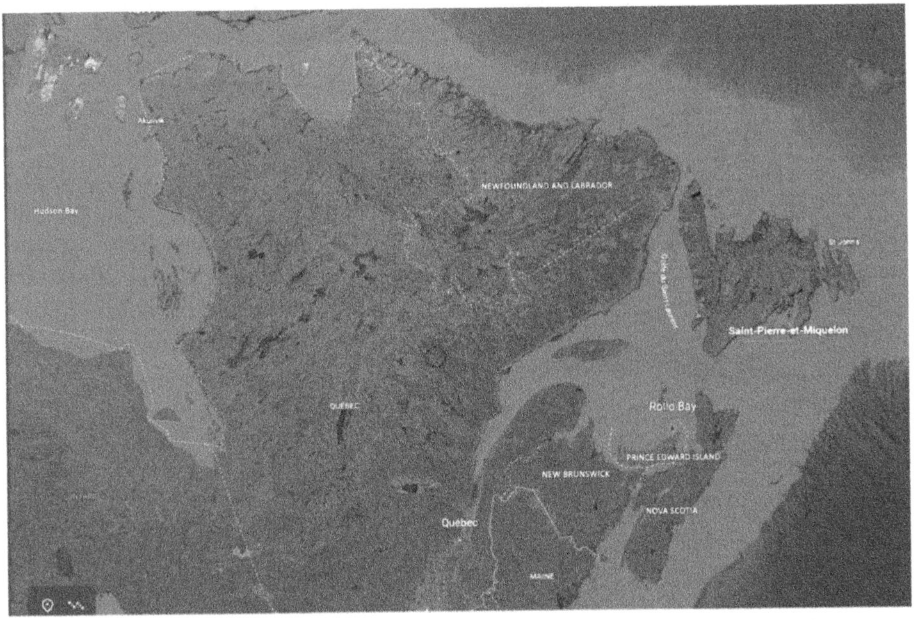

Figure 2.10. *Location of Rollo Bay (northern Canada), main aquaculture site for AquAdvantage transgenic salmon egg production (photo credit © Google Earth 2022)*

The grow-out phase of the transgenic juveniles was assigned to land-based ponds in Panama. This last rearing phase was intended to provide reassurance with regard to the eventual escape of the salmon, since they were not adapted to survive in a natural tropical environment.

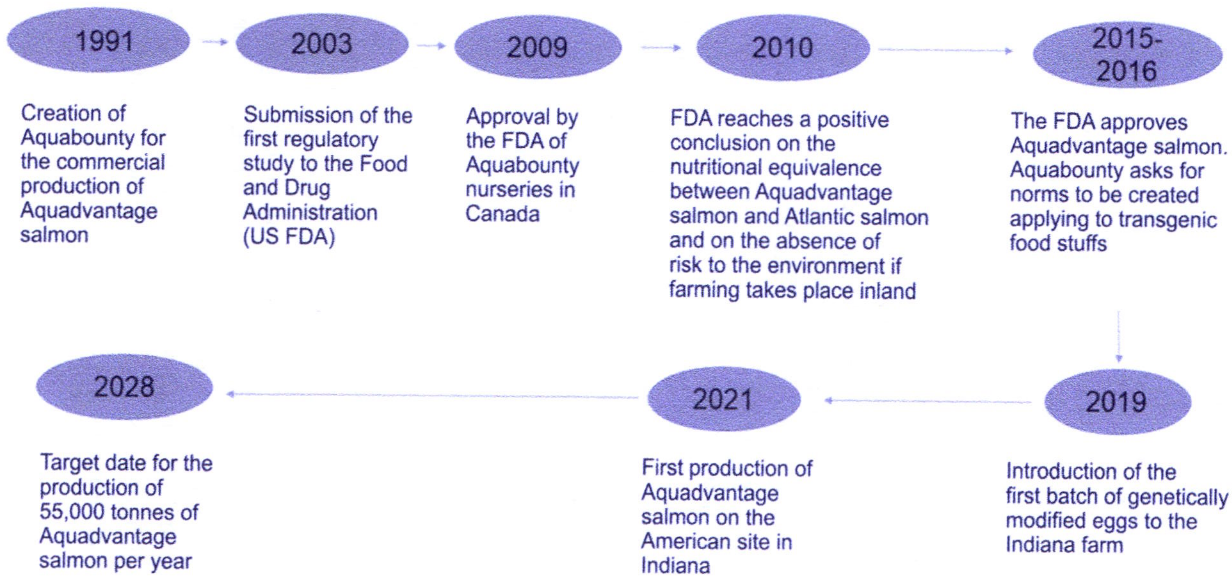

Figure 2.11. *Key stages in the development of AquAdvantage salmon production (from https://aquabounty.com/about-us)*

This experimental phase in Panama has now been superseded by the establishment of high-capacity farms at land-based sites in the Midwest (Albany, Indiana) and Canada (Rollo Bay, Prince Edward Island)[4]. The first site, which opened in June 2020, is capable of producing 1,200 tons of salmon a year. The Canadian site is announced as having a potential production capacity of 250 tons of fish per year. However, AquaBounty Technologies has set a production target of 55,000 tons of salmon per year by 2028 (see Figure 2.11).

In addition to the technological and production aspects, the development of the transgenic salmon aquaculture chain faced various regulatory and commercial constraints. Among the regulatory constraints imposed by the Food and Drug Administration (FDA) was the demonstration of nutritional equivalence between AquAdvantage salmon and Atlantic salmon. This demonstration was deemed satisfactory by the FDA. Another aspect raised by the FDA concerned environmental risk due to the possible escape of transgenic fish into the natural environment. This risk was considered to be minor in the case of land-based tank farming and with animals rendered sterile by triploidy. These two aspects, which are at the root of the FDA's authorization to market transgenetically engineered salmon, are still the subject of opposing arguments and controversy (see Table 2.3).

Approved in late 2015 in the United States and in 2016 in Canada, AquAdvantage salmon is the first transgenic fish authorized for food use.

In addition to regulatory constraints, transgenic salmon also faces a negative image with the public, which has an impact on its marketing. Described as "Frankenfish" by these opponents, this type of salmon is struggling to find its market niche, even in countries where it is authorized (the United States, Canada). In 2017, AquaBounty Technologies announced, however, that it had marketed 4.5 tons of AquAdvantage salmon in Canada to customers whose anonymity remains preserved. In 2021, the company published, in its annual preliminary report, a production of 84 tons of transgenic salmon over the first nine months of the year (GlobeNewswire 2021), but without indicating the buyers.

4 See: https://aquabounty.com/our-farms.

Domains	Arguments in favor	Arguments against
Nutritional equivalence	AquAdvantage salmon are biologically and chemically indistinguishable from Atlantic salmon (GlobeNewswire 2021)	AquAdvantage salmon muscle fibers are different from those of wild fish
Hormonal risk	AquAdvantage salmon produce the growth hormone like the wild type, with no increase in levels (GlobeNewswire 2021)	This hormone is produced all year round by transgenic fish, whereas its production is inhibited during the winter in wild fish
Health risks for livestock and consumers	AquAdvantage salmon are no more vulnerable to disease than Atlantic salmon. Rearing in land-based ponds would avoid the transmission of diseases present in the open environment (GlobeNewswire 2021). This situation would reduce the use of antibiotics, ensuring a lower risk for consumers (GlobeNewswire 2021)	These genetically homogeneous fish, produced outside the mechanisms of natural selection, could become more susceptible to disease more quickly
Risk of genetic pollution	Genetically modified eggs produce sterile females, and transgenic fish are reared in confined tanks. These conditions limit the risk of transmitting transgenes to wild salmon (GlobeNewswire 2021)	Triploidy sterilization is not 100% effective; 5% of eggs will produce fertile individuals

Table 2.3. *Partial summary of the advantages and disadvantages of producing transgenic salmon (from Wikipedia: https://fr.Wikipédia.org/wiki/AquAdvantage)*

In addition to Atlantic salmon, AquaBounty Technologies is focusing its development strategy on producing new transgenic species and obtaining marketing authorizations for them. The species targeted by this strategy correspond to those frequently produced by traditional fish farming, i.e. trout and tilapia. The Arctic char (*Salvelinus alpinus*) produced by aquaculture in Canada and Iceland is also one of the company's target species.

Aquaculture of transgenic fish has given rise to a new concept, dubbed the "blue revolution" by its promoters. It is presented today as an alternative production method to traditional fish farming.

AquAdvantage salmon is the result of genetic engineering operations initially on the transfer of genes responsible for coding for the production of antifreeze polypeptides and growth hormone, as detailed below.

2.1.2.1. *Transgenic salmon*

The production of transgenic salmon meets a need for aquaculture development in the northern regions of the American continent (the United States, Canada), which is limited, or even interrupted, during the winter period. During this period, the marine environment on Canada's coasts is characterized by the presence of ice and water temperatures below 0°C (0 to –1.8°C). In the case of salmon, a water temperature of -0.75°C is lethal for the animal.

The growth and survival of salmonids reared in floating cages are therefore severely affected by the extreme conditions observed in northern Canada during the winter. In particular, they have a negative impact on the economic viability of aquaculture farms located on Canada's northern coasts. However, some fish found in the North Atlantic (coasts of Labrador and Newfoundland, Iceland) are able to withstand temperatures close to the freezing point of seawater (–1.8°C) (see Figure 2.12). These include the wolffish (*Anarhichas lupus*) (see Figure 2.13), winter flounder (*Pseudopleuronectes americanus*) and ocean pout (*Trisopterus luscus*). These fish are characterized by the presence in their blood of high concentrations of polypeptides with colligative cryoprotective properties. Their protective effect seems to be due to their concentration and not to their biochemical nature. These polypeptides are known as antifreeze polypeptides (AFPs) (3,000–4,000 Da). They enable species equipped with them to tolerate the very low water temperatures of the North Atlantic and Arctic Oceans during the winter season.

Salmonids, and salmon in particular, lack this type of polypeptide and therefore do not benefit from the blood cryoprotection provided by AFPs. An initial experiment carried out on rainbow trout (*Salmo gairdneri* and *Oncorhynchus mykiss*) showed that the animal's resistance to cold was directly proportional to the quantity of purified AFPs taken from winter flounder injected into them (Fletcher et al. 1986). However, this approach requires an exogenous supply of AFPs for species that do not produce it. This observation led to the emergence of a new approach, namely the introduction into salmonids of genes responsible for coding for antifreeze polypeptides and under expression conditions enabling the production of AFPs sufficient to provide cold resistance in waters below 0°C.

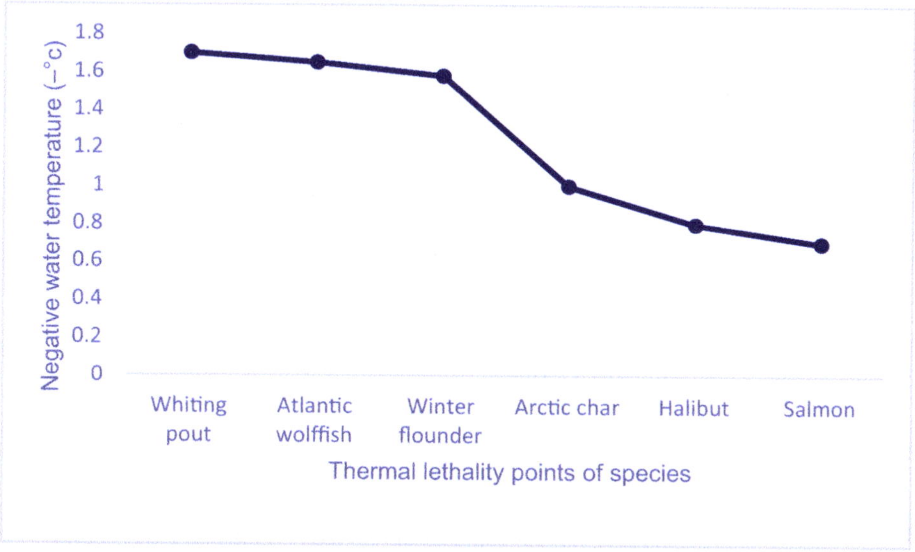

Figure 2.12. *Effect of water temperature on the lethality of different species (from Hew et al. (1995))*

In this context, the first experiments focused on Atlantic salmon, the main salmonid species produced by marine fish farming in North America. They involved the integration into the fish genome of genes encoding antifreeze polypeptides and growth hormone (Hew et al. 1995).

Figure 2.13. *Wolffish (*Anarhichas lupus*) with antifreeze polypeptides (photo credit © Citron 2012, Wikimedia/CC BY-SA 3.0)*

Fields of application	Expected benefits
Zootechnics	Improving the economic feasibility of fish farming Increased growth rates Improved feed conversion efficiency Use of low-cost diets Improved cold tolerance and frost resistance Improved fertility of breeding stock
Products and markets	Improved appearance Improved flesh color Improved aromas and flavors Improved texture Improved fatty acid composition
Biotechnology	Use of transgenic fish to produce molecules of pharmaceutical interest
Fundamental research	Helping to understand fish growth and reproduction through the use of model species (e.g. zebrafish and medaka)

Table 2.4. *Expected effects of transgenesis applied to fish on different production and application areas (from Fletcher and Davies (1991))*

These experiments led to the creation of AquAdvantage salmon, and represented the founding act of a new aquaculture based on the production of transgenic fish. Prior to this genetic engineering operation, the initiators of transgenesis in salmon described the expectations for this biotechnological revolution applied to aquaculture (Fletcher and Davies 1991). Among the expected benefits were improved growth rates and frost resistance (see Table 2.4).

Winter flounder produces an AFP called type I. This is a family composed of two major and several minor polypeptides. These polypeptides are encoded by a 1,000-base-pair AFP gene, which breaks down into a 600-base-pair intron sequence flanked by two exons. This gene has been selected for transfer to salmon.

The chosen AFP 2A-7 gene is then integrated into a linearized plasmid that has a cleavage site unique to the EcoR1 restriction enzyme. This 106-copy repeat genomic construct is then microinjected into fertilized salmon eggs via micropyle and left to incubate in water at 8°C.

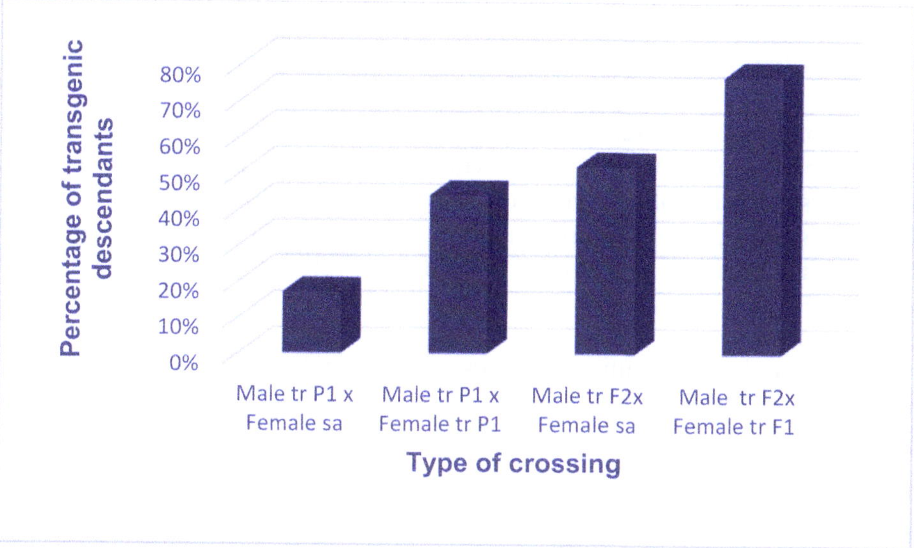

tr P1: parental transgenic; tr F1: first-generation transgenic; tr F2: second-generation transgenic; sa: wild type

Figure 2.14. *Impact of the type of cross between transgenic salmon and wild salmon on hereditary transmission of transgenes (from Hew et al. (1995))*

The presence of transgenes in the offspring of transgenic salmon has also been studied. The latter is linked to the type of crossing–breeding carried out. Indeed, the crossing of a transgenic male with a wild female leads to offspring of which 17% of individuals are endowed with transgenes capable of expression. A cross between a second-generation transgenic male and a first-generation transgenic female produces offspring in which 74% of individuals possess the transgenes (see Figure 2.14). The expression of the transgenes in these descendants could not be characterized.

This experiment shows that transgene transfer between genetically modified salmon and wild salmon is possible at the progeny level, although the latter remains less common. In addition, transmission between transgenic salmon is also possible, with high proportions of genetic inheritance of transgenes (72–74% of offspring). This type of result makes it possible to establish genetically stable lines of transgenic salmon. Nevertheless, the expression of the transgenes transmitted was not always determined during this study by the authors of this work.

Atlantic salmon are generally of commercial size after 14–16 months of rearing in the open sea (Hew et al. 1995). Increasing the animal's growth rate over a shorter rearing period is a zootechnical and economic challenge for salmonid aquaculture. In this context, numerous studies initially focused on the exogenous supply of growth hormones by injection in certain salmonids, such as coho salmon (*Oncorhynchus kisutch*) (Higgs et al. 1976). In this species, the intramuscular injection of bovine-derived growth hormones increases the animals' growth, as well as the food conversion ratio, particularly of proteins (Markert et al. 1977).

However, hormone injection does not appear to be an easy method to implement at farm level, and its commercial impact on the consumer may prove counter-productive.

In this context, promoters of transgenesis have developed a strategy for introducing the gene coding for growth hormone. Unlike poly AFPs, the concentration of growth hormone in fish blood is extremely low (50.10^{-9} $g.mL^{-1}$) (Hew et al. 1995). In this context, the overexpression of the gene introduced under physiological conditions represents a major challenge for successful transgenesis. In order to induce this overexpression, the gene to be introduced must be paired with a promoter that amplifies its expression. The use of promoters of viral or human origin has rapidly proved inappropriate in the context of developing foods for human consumption. The use of fish-derived promoters therefore became the preferred approach for this genetic engineering operation.

Promoters of AFP genes from fish were selected for this growth hormone gene transfer. A chimeric genetic construct integrating the AFP gene promoters and the GH gene was then developed. This construction, described as a "gene cassette" for all fish, was successfully tested on Atlantic salmon. Two possible constructions have been developed. The first is based on coupling the cDNA coding for growth hormone with the AFP gene promoter. This construct is called opAFP-GHc (growth hormone cDNA). The other construct is based on the association of the salmon-derived gene coding for growth hormone with the AFP gene promoter. This chimeric structure is called opAFP-GHf (growth hormone fish) (see Figure 2.15).

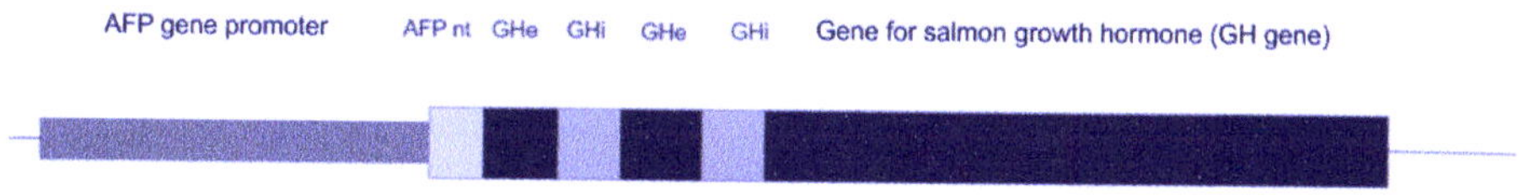

GHi: intron for the GH gene; GHe: exon for the GH gene; AFPnt: 5' sequence of the AFP promoter

Figure 2.15. *Example of a chimeric gene construct (opAFP-GHf) used for the transfer of the gene coding for growth hormone (GH gene) in Atlantic salmon (Salmo salar) (from Hew et al. (1995)). For a color version of this figure, see www.iste.co.uk/fleurence/aquatic.zip*

The injection of this type of chimera into Atlantic salmon eggs induces an effect on the animals' growth hormone levels, as well as on their daily body growth (see Figure 2.16). In transgenic animals, blood growth hormones are increased by a factor of 1.4 compared with non-transgenic fish. The most remarkable result is obtained for daily growth gain. Indeed, the latter can be at least four times higher for transgenic salmon if compared with non-transgenic animals (see Figure 2.16).

Both GHc and GHf constructs are effective in increasing the growth of transgenic salmon. Depending on the conditions, growth is improved by a factor of 4–10 (Hew et al. 1995). In some transgenic individuals, growth is even 10–30 times higher than that observed in non-genetically transformed animals. The phenotype and genotype of genetically modified fish also appear to be transmissible to offspring.

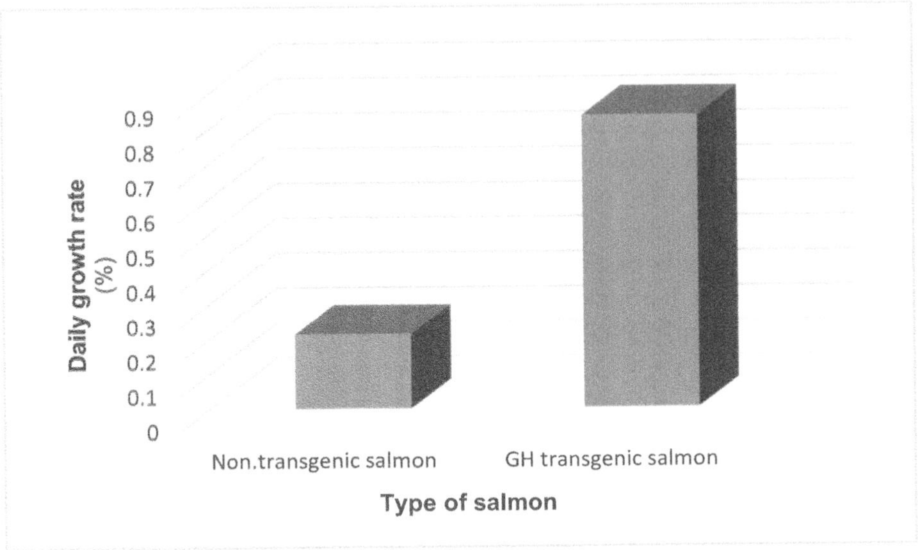

Figure 2.16. *Comparison of daily growth rates between non-transgenic salmon and transgenic salmon incorporating an opAFP-GHc chimeric construct (from Hew et al. (1995))*

The application of transgenesis to Atlantic salmon has prompted scientific and industrial promoters to coin a new concept known as the blue revolution. The blue revolution means developing an aquaculture industry based on the biological sciences and on engineering, capable of producing on a large scale and at low production costs in areas independent of proximity to the oceans, and under conditions that are less invasive for the environment. Improved growth rates, disease resistance, better feed conversion rates, shorter reproduction cycles and better recycling of water from the production system are all aspects of this blue revolution.

This new fish farming concept is based on the production of animals with zootechnical performance improved by transgenesis and reared in a confined environment in which rearing water is recycled in a closed system. The improved feed conversion rate of transgenic fish suggests that less nitrogenous waste would be discharged, resulting in less eutrophication of farm water. The blue revolution is therefore less impactful than traditional fish farming, which can be carried out in open, partially open or closed systems, but with a grow-out phase in the natural environment generating a risk of water pollution.

In addition to this environmental aspect, the promoters of this new aquaculture emphasize the lower costs of this type of production and therefore its ability to meet the food challenges of an overpopulated planet. However, the promoters of this type of aquaculture have not provided any information showing that it would be more profitable than traditional fish farming carried out under optimized conditions.

2.1.2.2. *Other transgenic species*

The injection of genes into fertilized eggs, a method used in Atlantic salmon, is also widely used for the production of transgenic fish of other species (Maclean and Laight 2000). A variant of this method involving microinjection of DNA into the pronucleus of unfertilized oocytes is also possible (Tanaka and Kinoshita 2001).

Some of these genetically modified species are used for basic research, while others are used for fish farming applications (see Table 2.5).

Species	Family	Common name
Oncorhynchus mykiss	Salmonidae	Rainbow trout
Oncorhynchus kisutch	Salmonidae	Coho salmon
Carassius auratus	Cyprinidae	Goldfish
*Danio rerio**	Cyprinidae	Zebrafish
Cyprinus carpio	Cyprinidae	Common carp
Oreochromis niloticus	Cichlidae	Tilapia
*Oryzias latipes**	Adrianichthyidae	Medaka
Ictalurus punctatus	Ictaluridae	Catfish or channel catfish

Table 2.5. *Examples of fish species genetically modified for fish production or for basic research applications (*) (from Maclean and Laight (2000))*

The species *Oryzias latipes*, known as medaka, is considered a model for the study of vertebrate development and physiology (see Figure 2.17). The modification of its genome by transgenesis allows us to characterize genomic regions whose transcription is important for morphogenesis or fish metabolism.

Figure 2.17. *Medaka (*Oryzias latipes*)* *(photo credit © Seotaro 2007, Wikimedia/CC BY-SA 3.0)*

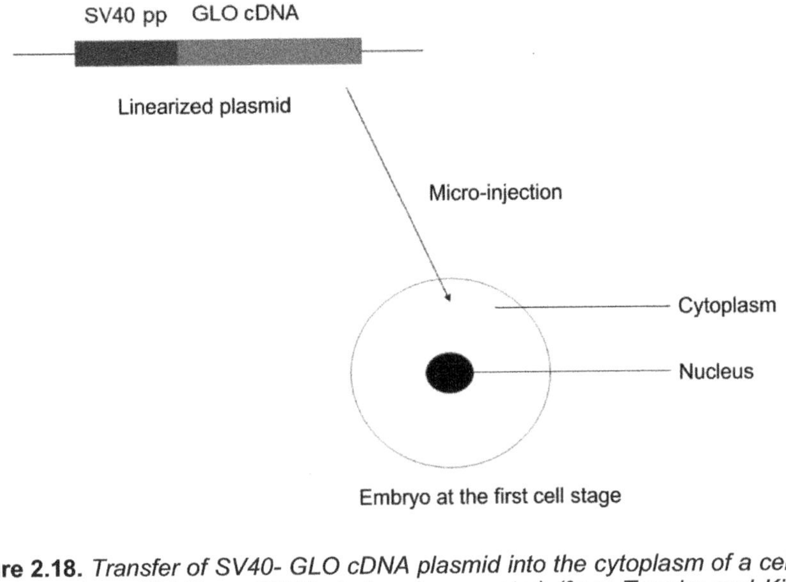

Figure 2.18. *Transfer of SV40- GLO cDNA plasmid into the cytoplasm of a cell-stage medaka embryo (SV40 pp: SV40 viral early promoter) (from Tanaka and Kinoshita (2001)). For a color version of this figure, see www.iste.co.uk/fleurence/aquatic.zip*

In the field of aquaculture production, transgenic medaka are used to understand the genetic mechanisms behind certain metabolic deficiencies and their correction. This is the case for vitamin C deficiency in fish and aquarium species such as trout, salmon, carp and guppy (*Poecilia reticulata*).

This type of deficiency can cause anemia, scoliosis and lordosis in animals if the dietary supply of vitamin C is insufficient. These species lack an enzyme called L-gulono-γ-lactone oxidase (GLO), which converts L-gulono-γ-lactone into ascorbic acid. This enzyme deficiency is due to an evolutionary mutation in the gene coding for the GLO protein. The correction of this mutation was tested on the model species medaka. The procedure used consisted of microinjecting (50.10^{-12} L) a linearized plasmid containing the cDNA of the gene encoding GLO coupled to an early promoter of viral origin SV40 into the cytoplasm of cell-stage medaka embryos (Toyohara et al. 1996; Tanaka and Kinoshita 2001) (see Figure 2.18). The survival rate of embryos one day after injection was 75%, with 82% giving birth to larva. After two months, four transgenic males reached sexual maturity, representing 23% of the biological resource initially transfected. The transgenic males with GLO activity were then crossed with non-transgenic females, in order to assess the hereditary transmission of the transgene. The transgene was found to be transmissible to offspring while retaining its expression. In this particular case, the medaka showed its usefulness as a study model for the development of a gene therapy potentially applicable to the production of fish species deficient in GLO activity.

The transgenic medaka is also a model organism for ecotoxicology research. As a sentinel organism, it can potentially be used to detect the presence of endocrine disruptors and mutagenic agents in the aquatic environment (Tanaka and Kinoshita 2001). To this end, the λ cII gene from the bacteriophage lambda virus infecting the bacterium *Escherichia coli* was transferred into the medaka (Winn et al. 2000). This gene, involved in the bacteriophage lysogenic cycle, was chosen because it is highly sensitive to exposure to mutagenic agents, and more particularly to N-ethyl-N-nitrosurea (ENU) (see Figure 2.19). This agent is a powerful mutagen used in particular to induce mutagenesis in gametes.

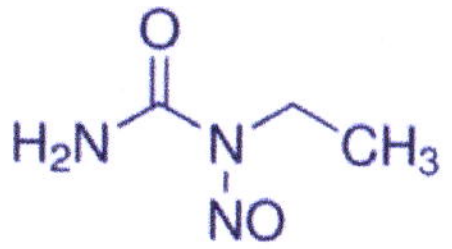

Figure 2.19. *Structure of the mutagenic molecule N-ethyl-N-nitrosurea (ENU)*

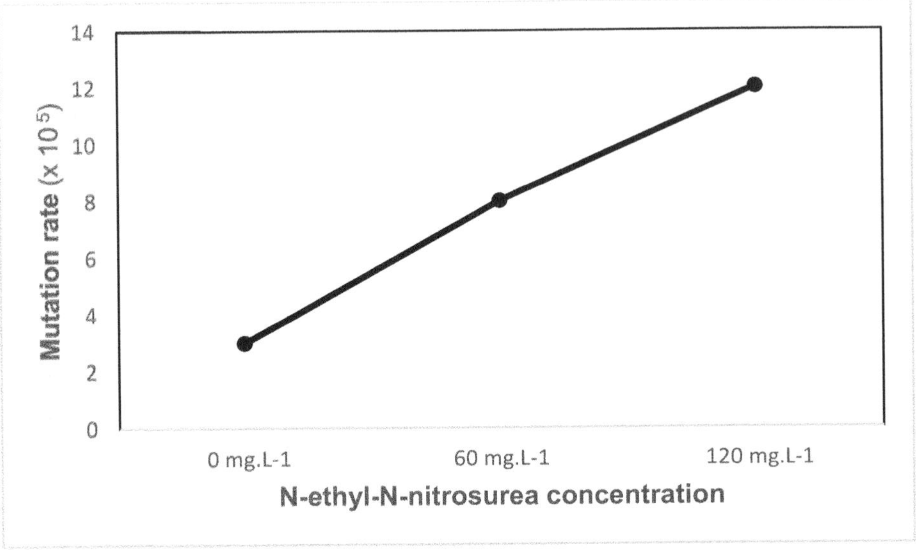

Figure 2.20. *Mutation rate in the λ cII gene introduced into transgenic medaka as a function of the dose of N-ethyl-N-nitrosurea (ENU) over a period of one hour (from Winn et al. (2000))*

In transgenic medaka, the mutagenic response is directly proportional to the concentration of ENU applied when the animal is exposed (see Figure 2.20). The mutation rate after exposure to the mutagen is three to four times higher than the spontaneous mutation frequency in non-exposed transgenic fish. This last point validates the use of transgenic medaka as a model species for mutagen detection. Moreover, the efficacy of this toxicological study model is comparable to that usually used for in vivo mutagenesis studies using transgenic rodents (Tanaka and Kinoshita 2001). In contrast, in the case of exposure to other carcinogens such as dimethylnitrosamine (DMN) (see Figure 2.21), transgenic λ cII medaka show a mutation frequency 20 times higher than that observed in transgenic λ cII rats treated with an equivalent dose (Hobbies et al. 2009).

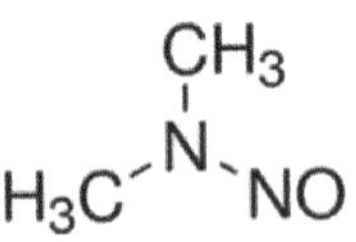

Figure 2.21. *Structure of the carcinogenic molecule dimethylnitrosamine (DMN)*

This model has also proved interesting for studying the hereditary transmission of genetic alterations produced by mutagens (Winn et al. 2008). The cII gene mutation observed after the exposure of transgenic males to ENU is found in postmeiotic germ cells (spermatids, spermatozoa). Males bred with non-transgenic females produce heterogeneous offspring in terms of mutation carriage. Three types of offspring are listed with a high, partial or even very low presence of the mutation in the offspring (see Figure 2.22).

When the cII gene mutation frequency is greater than 1.10^{-2}, it is assumed that the offspring has incorporated the mutated gene into all its body cells. The animal is then described as a full mutant.

When the mutation frequency detected is between 9.10^{-5} and $9.9\ 10^{-3}$, the offspring has partially integrated the cII transgene and the organism concerned is referred to as a mosaic mutant. Below these detection frequencies, the organism is considered non-mutant.

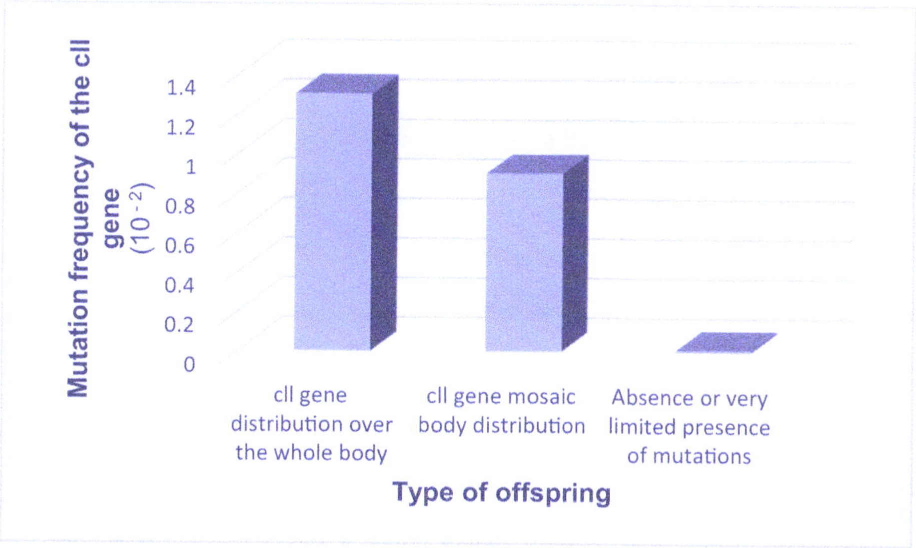

Figure 2.22. *Detection frequency for the cII gene mutation in the offspring of transgenic (cII) males exposed to ENU (from Winn et al. (2008))*

The transgenic ʎ cII medaka therefore appears to be a relevant model for studying germ cell mutagenesis following exposure to N-ethyl-N-nitrosurea. It enables us to assess the transmission of the transgene and its mutation to germinal cells. This approach also makes it possible to assess the rate of mutated germinal cells in the postmeiotic phase. The latter is generally 11%, higher than that of premeiotic germ cells (spermatogonia), which is limited to 3.5%. The medaka model also makes it possible to highlight the germinal cell targets affected by the transmission of the transgene. Indeed, the mosaic mutations seem to be the result of the mutation of the transgene affecting the spermatids, as well as the spermatogonia, which are germinal stem cells.

The use of the transgenic medaka model appears to be a fundamental research approach, enabling us to measure germ cell mutations and better appreciate the mechanisms of the hereditary transmission of genetic alterations resulting from mutagens. More generally, this model is extremely useful for assessing the toxicological impact of mutagenic pollutants on organisms and their genomes. As such, it is a particularly interesting tool for assessing environmental risks.

Figure 2.23. *Use of the transgenic OSP1P-GFP medaka to monitor the appearance of female germ cells (oocytes) in the animal's testes following ethinylestradiol exposure (from Zhao et al. (2014)) (photo credit © Seotaro 2007, Wikimedia/CC BY-SA 3.0)*

In addition to the effect of mutagens, transgenic medaka are also used to monitor the impact of endocrine disruptors on organisms, and in particular their sexual organs. In particular, they are used to monitor the appearance and evolution of intersexuality in the gonads of males exposed to the presence of female hormones (estrogens) in the environment (see Figure 2.23). This monitoring is carried out through the implantation of the GFP gene, which codes for the expression of a fluorescent protein coupled to the OSP1 gene, a sensitive and specific molecular marker of the onset of intersexuality in medaka males upon exposure to estrogens (Zhao et al. 2014). Among the estrogens most targeted by this type of study is ethinylestradiol, an estrogen used in the composition of many contraceptive drugs and found in the aquatic environment (see Figure 2.24).

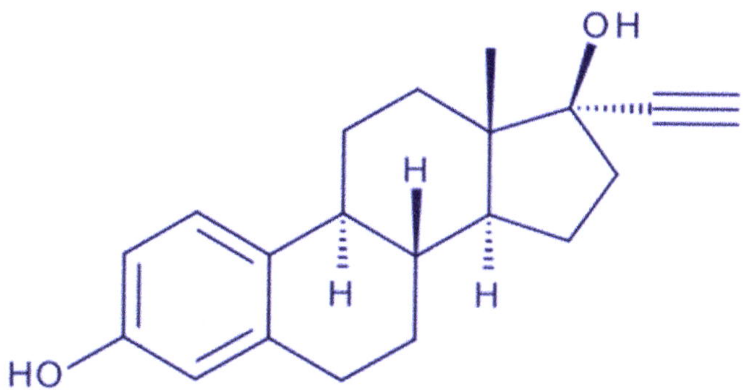

Figure 2.24. *Structure of the ethinylestradiol molecule estrogen, frequently used in contraceptive and gynecological drugs and likely to be found in watercourses*

The appearance of primary oocytes in the testes is observed early using GFP fluorescence. In particular, oocytes as small as 10 μm can be detected using this fluorescence technique, compared with 20 μm minimum oocytes with conventional histopathological analysis.

Similarly, the phenomenon of intersexuality is observable after 90 days of exposure to a dose of 0.75 ng.L^{-1} of ethinylestradiol, compared with a concentration of 1.38 ng.L^{-1} with histopathology.

In addition to the medaka, another transgenic fish species is also used for fundamental research. This is the zebrafish. This species (*Danio rerio*) belongs to the large Cyprinidae family, which includes the common carp. The search engine Google Scholar, which lists scientific articles, shows 221,000 references for a search based on the keywords "transgenic zebrafish", compared with 16,300 for "transgenic medaka".

In view of this, the transgenic zebrafish forms another model studied by the scientific community. As with the medaka, the zebrafish is widely used as a genetic model for the study of vertebrate development (Lin 2000).

It is also used as an in vivo model for studying inflammatory mechanisms, such as those affecting neutrophils (Renshaw et al. 2006). One of the promising areas associated with the use of transgenic zebrafish remains the study and understanding of the mechanisms behind neurodegenerative diseases such as Parkinson's and Alzheimer's (Sager et al. 2010).

Indeed, zebrafish are characterized by the presence of genes shared with the human species, which are known to be involved in the main degenerative disorders of the central nervous system, Parkinson's and Alzheimer's diseases.

These so-called orthologous genes are common to both species and have retained a similar structure and function throughout evolution (see Table 2.6).

Furthermore, the central nervous system of zebrafish is organized in the same way as that of other vertebrates and is therefore similar to that of humans. In this sense, zebrafish neuroanatomy appears to be a relevant model for the study of human neurodegenerative diseases.

Assessment of the vulnerability and activity of neuronal groups on the neural axis, as well as the expression of genes encoding different neuronal classes, can be visualized using genetically modified zebrafish. To do this, the fish must have integrated the gene coding for the fluorescent GFP protein coupled with promoters or genes playing a role in neuronal development (GATA2, HuC gene, ENO2). The approach based on the genetic coupling of GFP and ENO2 provides a useful model for the study of tauopathy. Tauopathy, an umbrella term for some 20 pathologies characterized by a lack of understanding of the role of the Tau protein, remains a priority for understanding neurodegenerative diseases.

Diseases	Human gene	Zebrafish gene	Protein	Protein length (in amino acids)	Composition in amino acids (% similarity)
Parkinson's	Dj-1	*dj-1*	DJ-1	189 (human and zebrafish)	89
	PRKN	*prkn*	Parkin	465 (human) 458 (zebrafish)	75
Alzheimer's	APP	*appa*	Amyloid precursor protein	695 (human) 738 (zebrafish)	74

Table 2.6. *Examples of orthologous gene encoding proteins involved in or associated with degenerative diseases such as Parkinson's and Alzheimer's (from Sager et al. (2010))*

Transgenic zebrafish can also be used as a biological sentinel for aquatic pollutants (Carvan III et al. 2000). Detectable pollutants include aromatic hydrocarbons, polycyclic hydrocarbons and various halogenated molecules, such as dioxins and polychlorinated biphenyls. Other pollutants such as heavy metals or estrogens can also be detected, depending on the motifs or DNA sequences introduced into the pollutant-response genetic system (see Table 2.7).

More specifically, the transgenic zebrafish incorporates a gene (RP) that is activated when pollutants accumulate in the animal's tissues. This gene is coupled to a gene coding for luciferase (Luc gene), the regulation of which is dependent on the activity of the RP gene. Activation of the RP gene induces the expression of the Luc gene and hence the production of luciferase. This enzyme then produces a detectable and measurable bioluminescence reaction, indicative of the presence of pollutants in the environment (see Figure 2.25). The bioluminescence reaction is easily detected in zebrafish cell culture, as well as intact, live animals. This latter aspect is an important asset in the development of in vivo and in situ methods for assessing pollutants in the aquatic environment.

Response types	DNA sequence 5'–3'	Activating pollutants
Aromatic hydrocarbons	TWGCGTG	Polycyclic aromatic hydrocarbons Dibenzofurans Dioxins
Heavy metals	TGCRNCGG	Heavy metals (Pb, Hg, Cd, etc.)
Estrogen	GGTCANNNTGACC	Pharmaceutical estrogens, pesticides, phytoestrogens, etc.
Retinoids	GGGGTCAAAGGTCA GGGGTCATGGGGTCA	Retinoic acid and natural or pharmaceutical derivatives

Table 2.7. *Examples of DNA sequences or motifs used to develop a specific response gene to pollutants of different kinds (N; A, T, G or C; R: A or G; W: A or T) (from Carvan III et al. (2000))*

Transgenic zebrafish also bioconcentrate pollutants present in the environment, facilitating their detection. For example, dioxin diluted to a concentration of 10^{-17} M in the aquatic environment is found at a concentration of 10^{-12} M in the tissues of transgenic fish. The pollutant is thus concentrated 100,000 times over its initial concentration, making the transgenic zebrafish a powerful tool for detecting pollutants that are naturally highly diluted in the natural environment. From a theoretical point of view, this effect of bioaccumulation of pollutants by zebrafish makes it a potential candidate for environmental depollution. However, given the ecological risks involved, this would appear to be neither feasible nor desirable in a natural environment. On the other hand, in a strictly confined environment, this hypothesis deserves to be discussed and validated by in-depth research.

The transgenic medaka and zebrafish mainly appear as study models for fundamental research in the fields of developmental biology, certain human diseases and ecotoxicology. They now play the role that laboratory rodents and more particularly mice have long filled. Easier and more ethical to use than small mammals, they are gradually establishing themselves in the scientific community as useful animal models for in vivo or in situ studies.

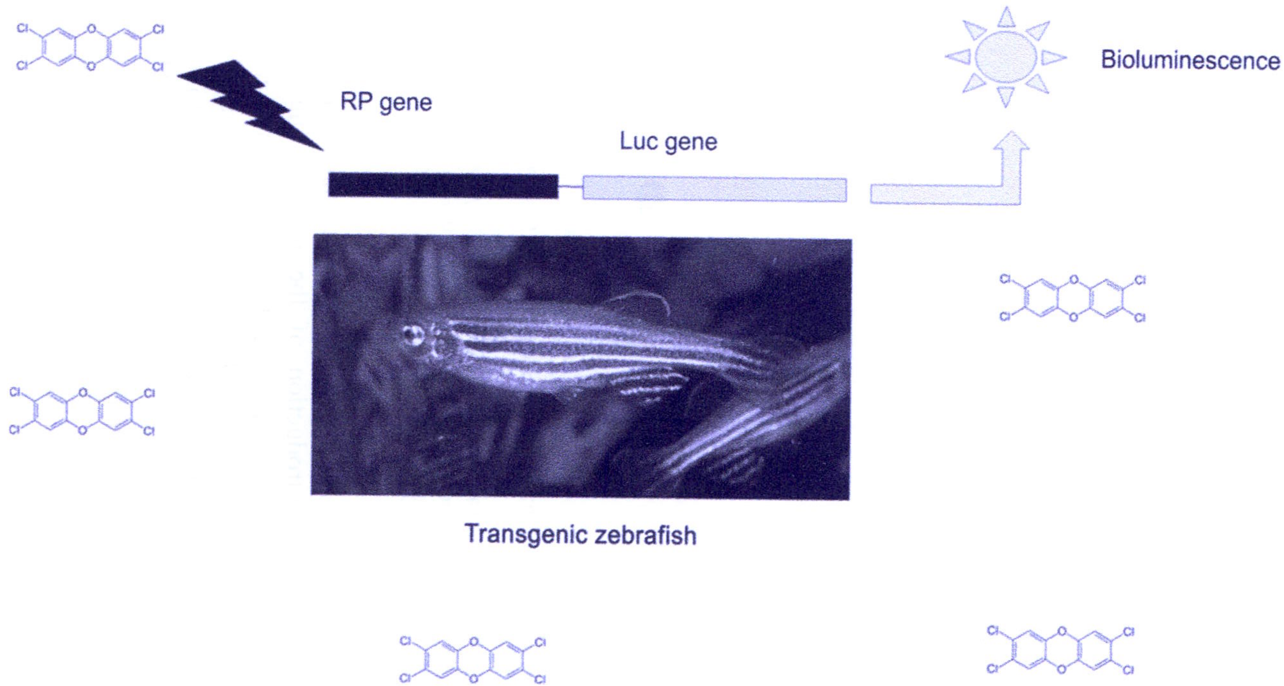

Figure 2.25. *Zebrafish incorporating a pollutant-response gene (RP gene) coupled to the to the inducible luciferase gene (Luc gene) used as a biological sentinel in a dioxin-contaminated environment (from Carvan III et al. (2000)) (photo credit © Azul, Wikimedia Commons). For a color version of this figure, see www.iste.co.uk/fleurence/aquatic.zip*

Apart from this application framework, these species are mainly used in aquariophily and are therefore not among the main species produced by fish farming for food purposes.

In addition to transgenic Atlantic salmon, other species of interest to fish breeders are also being transgenicized. These include carp and tilapia, which, like salmonids, are flagship species in world fish culture.

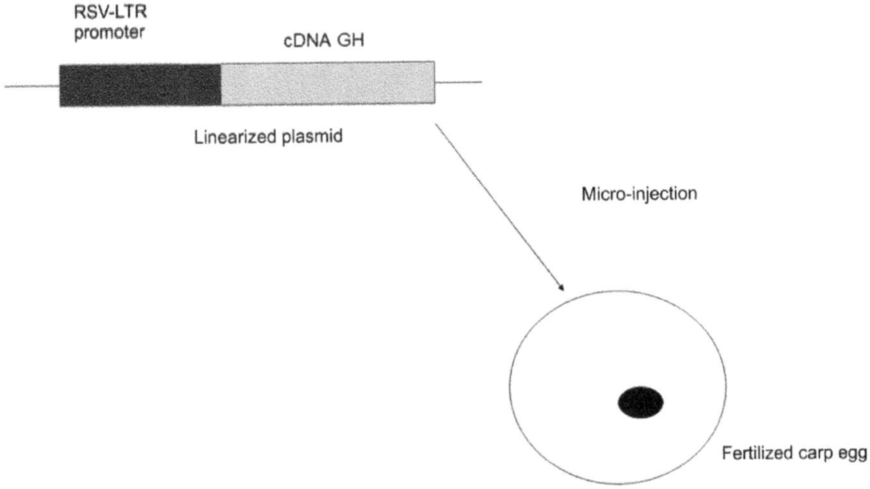

RSVL-LTR: sequence of viral origin; cDNA GH sequence from *Oncorhynchus mykiss*.

Figure 2.26. *Process for transferring cDNA coding for growth hormone in a fertilized carp egg by microinjection (from Zhang et al. (1990)). For a color version of this figure, see www.iste.co.uk/fleurence/aquatic.zip*

The common carp (*Cyprinus carpio*) has been the subject of transgenesis experiments involving the introduction of the gene coding for growth hormone (GH). As with transgenic Atlantic salmon, this approach is designed to improve fish production. The DNA sequence transferred is a cDNA-encoding growth hormone derived from the rainbow trout GH gene (*Oncorhynchus mykiss*) (Chatakondi et al. 1995) (see Figure 2.26). The recombinant structure is a linearized plasmid consisting of a viral promoter (RSV-LTR) and the GH cDNA. This plasmid is microinjected into a fertilized egg cell of the target species. Transgenic individuals are then bred to produce a line of first-generation (F1) transgenic carp.

The aim of this approach is to accelerate the carp's growth process, as has been done for Atlantic salmon. However, the study implemented in 1995 based on this approach presents another challenge in terms of applications. The aim was to evaluate the influence of transgenesis on the body composition of carp and thus the impact of this operation on the quality of the product offered to the consumer. The body protein content of transgenic fish was significantly higher than that of non-transgenic carp (+7.5%) (see Figure 2.27). Conversely, the water and body fat contents of transgenic fish were also different from those recorded for non-genetically modified fish. In addition to these results, amino acid composition, a key determinant of the nutritional value of dietary proteins, also differs according to the type of fish (see Figure 2.28).

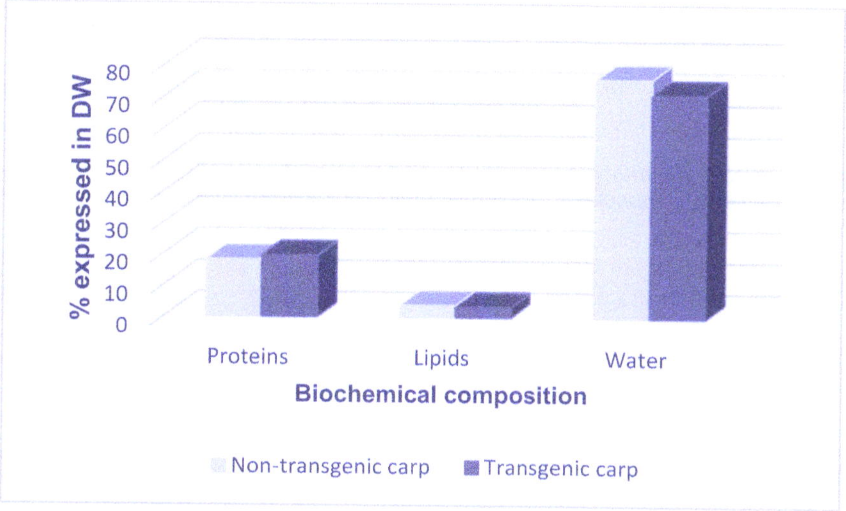

Figure 2.27. *Body biochemical composition of transgenic carp and non-transgenic carp (from Chatakondi et al. (1995))*

This carp experiment shows that the expression of the GH gene from rainbow trout has an effect on the biochemical composition of carp carcasses. This results in an increase in protein content in the transgenic animal, as well as a change in the amino acid profile. The most striking effect is on fat content (see Figure 2.29). This is reduced by 13% in the case of transgenic carp. On the other hand, the dietary energy value calculated according to the protein and lipid composition of the flesh shows no significant difference between transgenic fish (113.13 Kcal/100 g) and non-genetically modified fish (111.71 Kcal/100 g).

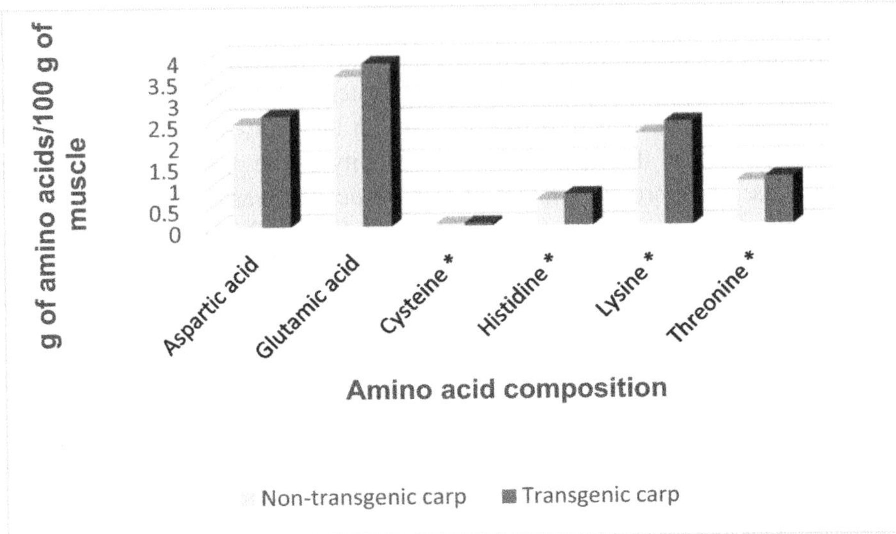

Figure 2.28. *Significantly different amino acid compositions between transgenic and non-transgenic carp muscles (*: essential amino acid) (from Chatakondi et al. (1995))*

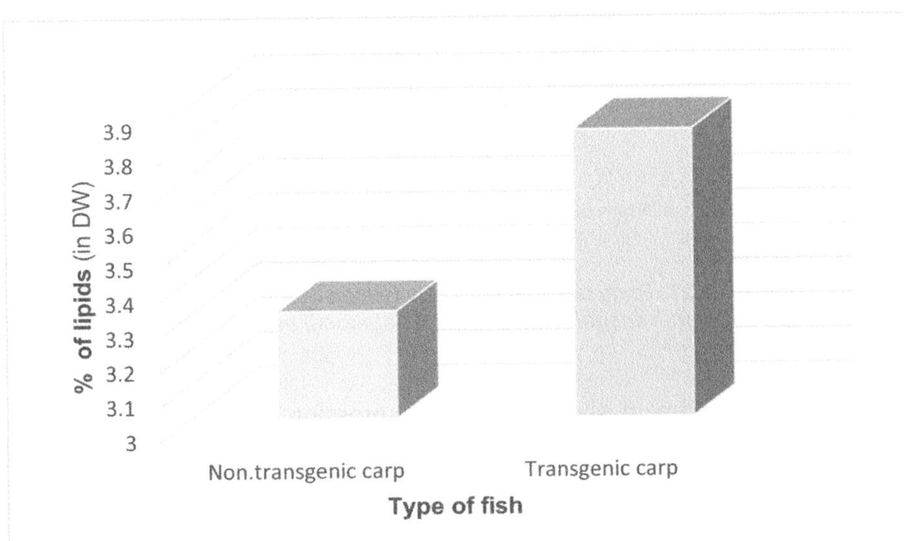

Figure 2.29. *Lipid composition in the carcasses of transgenic carp and non-transgenic carp (from Chatakondi et al. (1995))*

Increasing protein content appears to be a beneficial criterion for human consumption. Similarly, lower fat content is a factor in reducing the risk of rancidity and increasing food shelf life. However, these positive criteria for product quality need to be considered in the light of the potential ecosystem impacts that may be generated by the production of transgenic carp, the first species to be reared mainly in open or partially open environments.

A comparable study carried out on transgenic carp incorporating a gene coding for growth hormone but of human origin gave similar results (Fu et al. 1998). Regardless of the diet provided, transgenic fish were characterized by a higher body mass and protein content (+5%) than control fish. Their lipid content was also lower (-12%). On a low-protein diet (20%), transgenic carp showed a higher palatability for the protein ration provided than that recorded with control carp. On the other hand, fecal excretion of proteins was lower with transgenic animals. This appears to be linked to the digestive process, which is significantly stronger in transgenic carp. An improvement in the apparent digestibility of the feed ration is in fact observed for transgenic animals (+1.82%). These differences in feed intake and nutritional physiology between the two types of fish are no longer observable when the animals are subjected to a higher-protein diet (40% protein). The reduction in the amount of nitrogenous waste discharged into the environment following improvements in the protein conversion factor and nutritional physiology is an argument taken up by the designers of the blue revolution.

However, these studies on transgenic carp show that there is no biochemical and therefore nutritional equivalence between the genetically modified fish and the non-genetically modified fish. This kind of argument would therefore be considered favorable by the FDA in a request to authorize this transgenic fish in the USA.

The production of transgenic fish advocated by proponents of the blue revolution also comes up against the risk of transgenic animals escaping and the emergence of genetic pollution in natural ecosystems.

A scenario known as the "Trojan gene" lays the foundations for such a risk (Muir and Howard 1999). This scenario is based on the voluntary escape of a transgenic male transformed with the gene coding for growth hormone and its contact with a wild population. The hypothesis put forward is that the transgenic male, larger than the wild male, would be preferentially selected by wild females. In this hypothesis, the transgenic male would therefore have a reproductive advantage. However, this reproductive advantage could be counterbalanced by reduced adaptation of the transgenic fish to the natural environment and therefore reduced life expectancy.

This scenario was tested with the medaka model, in which a genetic chimera coding for the growth gene was introduced. This chimera took the human gene coding for growth hormone (GHh) coupled to a promoter of the growth gene, albeit obtained from salmon (GHsp). With a high mating advantage (4.7), the introduction of a transgenic male into a wild population could, depending on the viability of his offspring, lead to the extinction of the wild population after some 30 generations (scenario three) (see Figure 2.30).

The reproductive advantage of the transgenic medaka, although determined, is at all times modulated by the viability of the offspring. Despite this, a strong mating advantage associated with moderate viability (0.45 < viability < 0.9) could induce a significant risk of extinction at the wild population level.

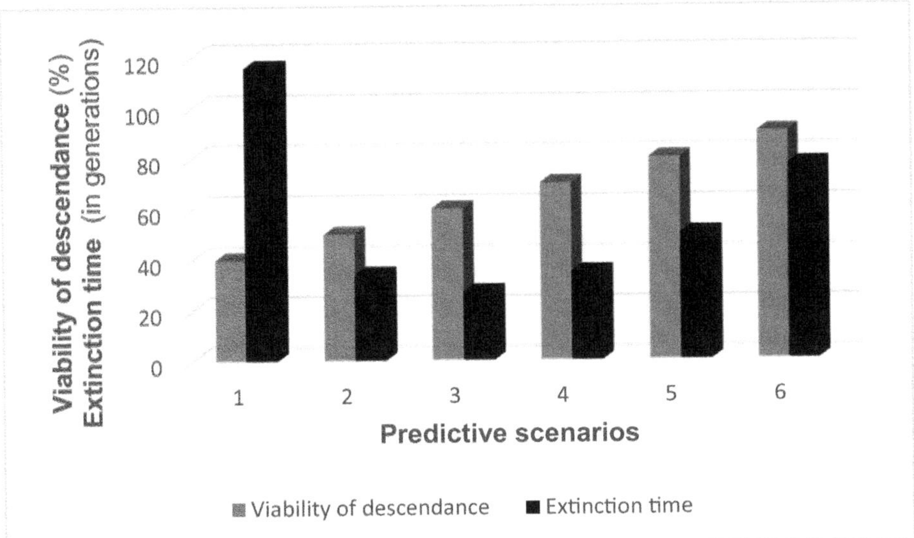

Viability: 1: 40%; 2: 50%; 3: 60%; 4: 70%; 5: 80%; 6: 90%

Figure 2.30. *Influence of different predictive scenarios on the extinction time of the wild medaka population compared to the viability of transgenic offspring under a strong mating advantage (4.7) (from Muir and Howard (1999)). For a color version of this figure, see www.iste.co.uk/fleurence/aquatic.zip*

This predictive model also suggests that the introduction of a transgenic male with high mating ability and generating offspring with high viability could in the very short term cause the extinction of the corresponding wild population.

The reproductive advantage of a transgenic fish introduced accidentally or deliberately into a natural ecosystem remains one of the major risks associated with the production of this type of animal.

To mitigate this risk, a sterilization approach can be implemented to prevent the reproduction of animals undergoing transgenesis or breeding.

Several methods can be used to achieve this goal. The most common is sterilization by triploidy induction. The acquisition of triploidy renders organisms sterile. The animal's cells are no longer endowed with chromosome pairs (2N), but with chromosome triplets (3N). Chromosome pairing in the form of triplets prevents meiosis from proceeding normally and so prevents the normal production of haploid gametes (N), thus inhibiting the reproductive process.

Triploidy can be induced by thermal shock. This process has been widely applied to fish species, including Atlantic salmon (Johnstone 1985). For this species, heat shock is generally applied to recently fertilized eggs, i.e. 20 minutes after fertilization. The heat stress temperatures can vary between 26 and 32°C over a period of 4–20 minutes. At 30°C, egg survival decreases with an increasing duration of heat shock (see Figure 2.31). On the other hand, increasing the duration of the shock improves the triploidy rate.

For aquaculturists, the choice is between an acceptable mortality rate and efficient triploidy. The consensus sought in the case of non-genetically modified salmon is quite different than for transgenetically modified salmon.

Triploidy is a sterilization tool frequently used with Atlantic salmon (*Salmo salar*) in order to limit the risks to wild populations in the event of individuals escaping from breeding pens (Benfey 2001, 2016). This operation makes it possible to produce triploid, and therefore formally sterile, females with an efficiency rate of over 98%. Initially based on thermal shock, the current process relies mainly on the physical process of pressurizing fertilized eggs. The pressures applied vary but, for Atlantic salmon, the standard pressure is 65,500 kPa for 5 minutes (Benfey 2016).

AquaBounty Technologies, a promoter of transgenic salmon, has announced a 99.5% increase in triploid placement efficiency using the hydrostatic pressure process.

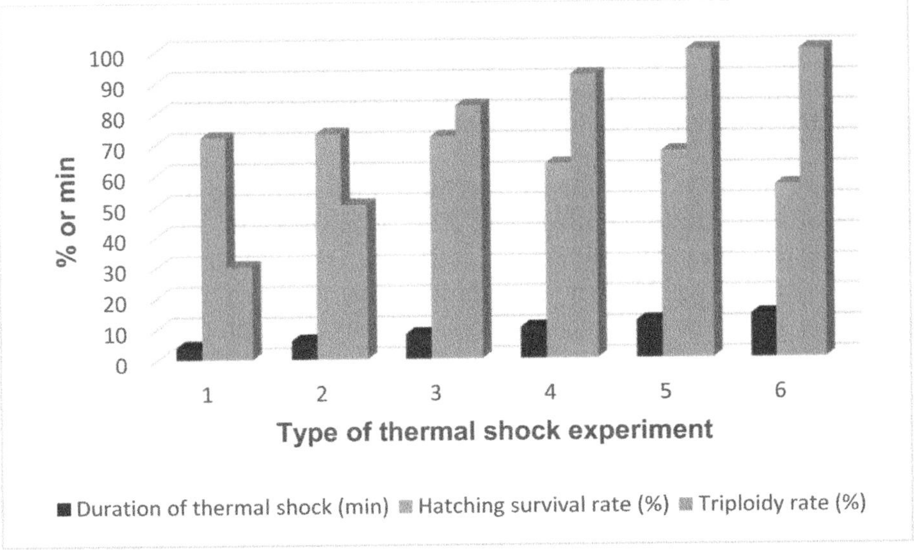

1: 4 minutes; 2: 6 minutes; 3: 8 minutes; 4: 10 minutes; 5: 12 minutes; 6: 14 minutes

Figure 2.31. *Time-dependent effect of a 30°C heat shock on survival and triploidy of fertilized salmon eggs (from Johnstone (1985)). For a color version of this figure, see www.iste.co.uk/fleurence/aquatic.zip*

The exceptional yields described are, however, moderated by those put forward by other sources. For example, Houdebine (2007) reports that triploidy-generating processes are approximately 95% efficient. AquaBounty Technologies also admits that 5% of treated eggs remain fertile. Out of a currently announced production of 15 million transgenic eggs, this would potentially generate 750,000 fertile transgenic fish.

Apart from triploidy, other methods of sterilization are possible for salmonids. These include the hormonal treatment of fry or juveniles. This treatment relies mainly on the use of a male hormone, 17 alpha-methyltestosterone (see Figure 2.32). It is generally coupled with the use of another hormone, which may be 3,5,3'-triiodo-L-thyronine (see Figure 2.33) (Shelbourn et al. 1992).

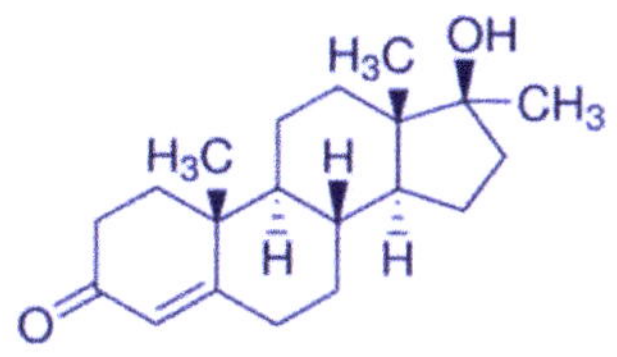

Figure 2.32. *Structure of 17 alpha-methyltestosterone*

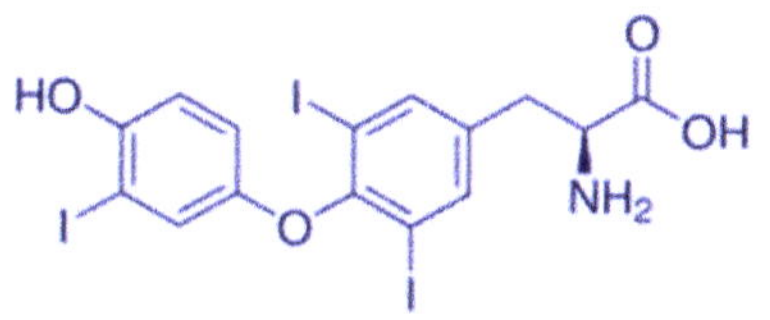

Figure 2.33. *Structure of 3,5,3'-triiodo-L-thyronine*

In coho salmon (*Oncorhynchus kisutch*), methyltestosterone (MT) is administered by immersing fry in a solution containing 200 mg of hormone per liter. Hormone treatment is then provided via the diet, at a rate of 10 mg MT per kg ration. This completes the first phase of the sterilization process. After 12 weeks of methyltestosterone food, a second phase based on 3,5,3'-triiodo-L-thyronine is implemented. A third phase involving the supply of MT at variable concentrations and transfer to water at 11°C finalizes the sterilization process (Shelbourn et al. 1992).

This process produces sterile juveniles with a 97% survival rate, but with a lower weight than non-sterilized fish (-33%) and a lower ability to convert the feed supplied (-43%).

The sterilization process used on unmodified coho salmon is potentially applicable to transgenic coho salmon.

However, the hormonal action of MT can have unwanted side effects in salmonids. For example, when used with 17 β estradiol, it can inhibit smoltification (see Figure 2.34) (Miwa and Inui 1986). This phenomenon is essential for salmonids (smolts), as it is at the origin of the adaptation of anadromous fish to the transition from fresh to seawater.

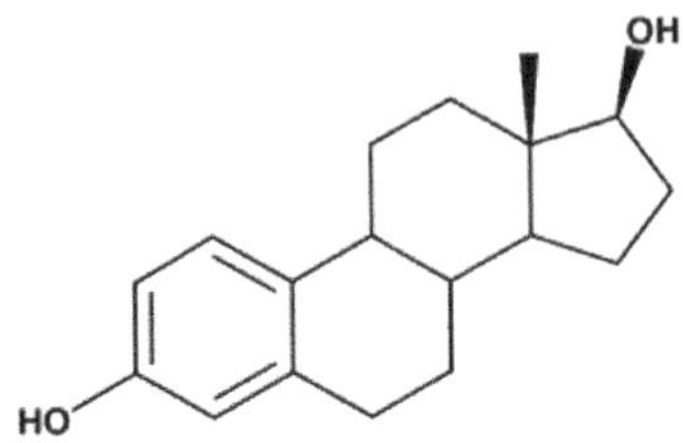

Figure 2.34. *Structure of 17 β-estradiol*

Research on amago salmon (*Oncorhynchus rhodurus*) shows that, in this species, the inhibitory effect of the MT hormone is far greater than that of estradiol.

Hormonal sterilization methods are therefore tricky to implement, and need to be adapted to each species. Developed on non-transgenic animals, they can be applied to transgenic fish. However, they have the major disadvantage of introducing endocrine-disrupting substances directly or indirectly into the environment. For this reason, their use in fish production for food purposes is not insignificant.

In addition to physical (triploidy) or chemical (hormonal) sterilization, there is another alternative approach applicable to transgenic fish. This approach, known as “genetic sterilization”, is itself based on transgenesis. It involves modifying the expression of a gene or genes required for fertility within the organism.

This can be made possible by transferring antisense genes (synthetic DNA sequences) which produce RNAs likely to interfere with the translation of a protein, such as gonadotropin, the gonadotropic hormone that plays a fundamental role in vertebrate, and therefore fish, reproduction (see Figure 2.35).

Regardless of sterilization methods, in order to limit the risk of genetic pollution of wild populations by transgenic fish, the geographical isolation of fish farms remains an option. This approach is based on closed-system aquaculture, far from coasts or waterways. This last aspect represents one of the pillars of the blue revolution advocated by the promoters of AquAdvantage salmon. In their view, transgenic fish should be produced in land-based ponds in a confined environment. Moreover, some authors suggest rearing transgenic fish in natural, geographically isolated locations (Maclean and Laight 2000). Finland, a sparsely populated country with an abundance of isolated lakes, was mentioned as a possible site for the rearing of transgenic alpine trout (*Salvelinus alpinus*) by Maclean and Laight (2000) in their review of the risks and benefits associated with the production of transgenetically engineered fish.

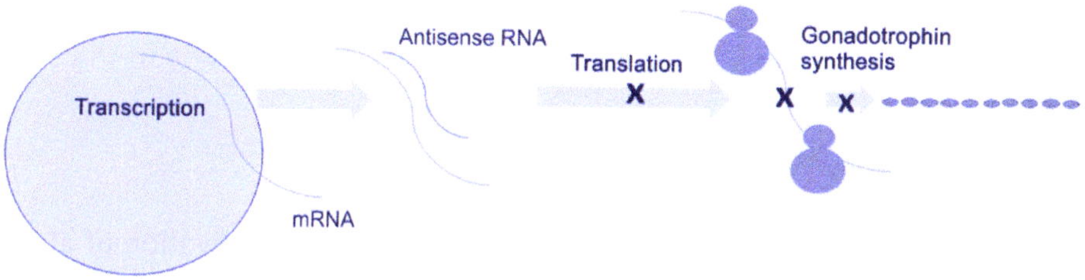

Figure 2.35. *Mode of action of an antisense RNA inhibiting the translation of mRNA coding for gonadotropin hormone synthesis in fish (from Maclean and Laight (2000))*

Transgenic fish are based mainly on the *Salmo salar* species. Other salmonid species are being researched as part of the development of the blue revolution concept. This new mode of fish production is mainly Western and calls on highly advanced aquaculture technology in the fields of genetic engineering and plant engineering. At the same time, it faces strong regulatory constraints in the United States and Canada, the countries promoting this aquaculture revolution. Its development in the European community is severely limited by the ban on producing and marketing this type of fish (see section 5.1).

Carp is one of the most widely farmed fish species. Since the late 1980s, the common carp has been genetically modified for experimental purposes. This species, like the transgenic tilapia, is not currently affected by the blue revolution. This is probably due to the low economic value added by this type of species compared with that generated by salmonids.

Quite apart from the fish-farming aspects and the resulting food applications, transgenesis is an approach that is proving its worth in the field of fundamental research when applied to reproductive biology and the development of vertebrates. This approach, based on the use of transgenic fish, also appears to be an interesting tool for assessing the risks generated by certain anthropogenic pollutants in aquatic ecosystems.

The use of transgenic fish is therefore subject to two very different purposes, namely the production of animals for food purposes and the development of organisms dedicated to research, whether in fundamental biology or ecotoxicology.

2.2. Algoculture

Algoculture, also known as phycoculture, is the mass cultivation of algae. This activity can involve macroscopic or macroalgae as well as microscopic or microalgae.

2.2.1. *Traditional seaweed farming*

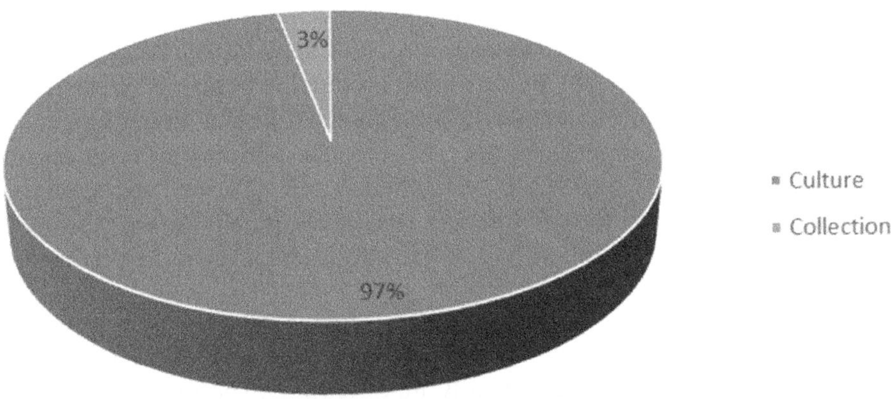

Figure 2.36. *Distribution of aquaculture activity and harvesting in the global production of algal resources (according to FAO (2022a)). For a color version of this figure, see www.iste.co.uk/fleurence/aquatic.zip*

Traditional seaweed farming can be defined as the cultivation of genetically unmodified algae. It covers both macro- and microalgae. The world production of algae was estimated to be 37 million tons in 2022, expressed in fresh weight (FAO 2022a). This production is almost exclusively due to aquacultivation. Today, harvesting accounts for 3% of bioalgal mass production compared with 10% some 15 years ago (see Figure 2.36).

Worldwide production of algae for food purposes is mainly based on macroalgae. Macroalgae can be used directly or indirectly for human consumption. Directly, macroalgae are consumed as sea vegetables, ingredients or intermediate food products (IFPs) (see Figure 2.37).

Figure 2.37. *Examples of the use of wakame seaweed (*Undaria pinnatifida*) and nori (*Porphyra yezoensis*) as vegetable side dishes (photo credit © Fleurence J. 2018)*

Indirectly, they are exploited for the extraction of phycocolloids (alginates, agar, carrageenans) used by the food industry as additives.

The main species cultivated worldwide for these two types of applications belong to the genus *Laminaria*. Within this genus, the most widely produced species is *L. japonica*, known to the general public as kombu (or konbu). *Laminaria japonica* or *Saccharina japonica* is mainly farmed in China, the world's leading supplier (see Figure 2.38). Production of *L. japonica* has soared over the last decade (see Figure 2.39), reaching almost 12 million tons (fresh weight) in 2018 (FAO 2020).

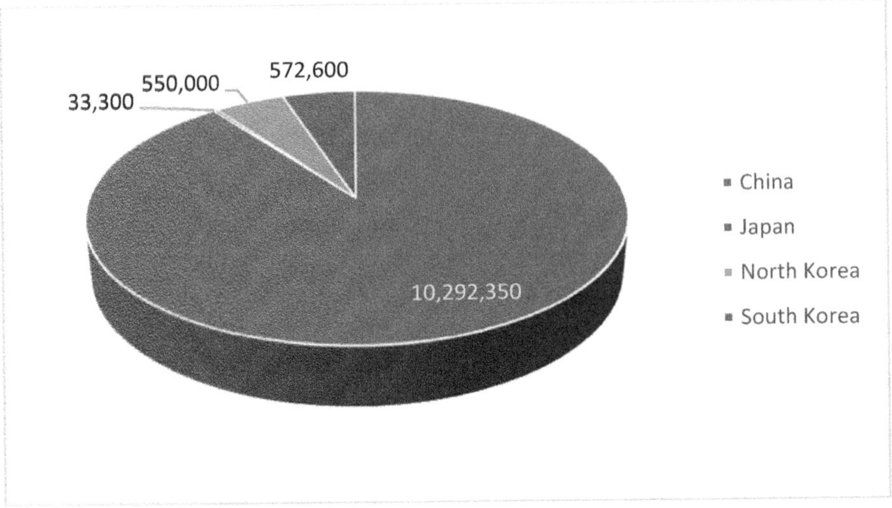

Figure 2.38. *Worldwide distribution of* Laminaria japonica *production (in tons fresh weight) by country (according to FAO (2020)). For a color version of this figure, see www.iste.co.uk/fleurence/aquatic.zip*

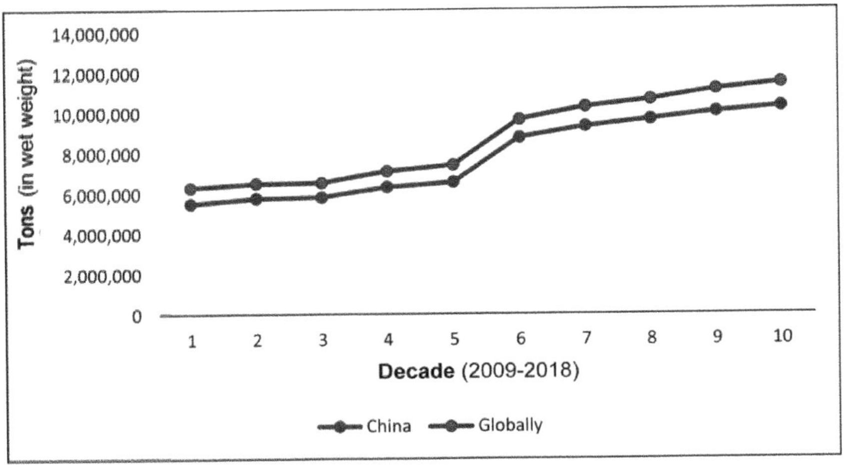

1: 2009; 2: 2010; 3: 2011; 4: 2012; 5: 2013; 6: 2014; 7: 2015; 8: 2016; 9: 2017; 10: 2018

Figure 2.39. Laminaria japonica *(*Saccharina japonica*) production trends worldwide and in China between the years 2009 and 2018 inclusive (according to FAO (2020)). For a color version of this figure, see www.iste.co.uk/fleurence/aquatic.zip*

The development of *L. japonica* and laminariales is based on a heteromorphic digeneous cycle. This results in the alternation of two generations, one diploid (2n), known as the sporophyte, and the other haploid (n), known as the gametophyte. The sporophyte is macroscopic, while the gametophyte is microscopic, hence the term "heteromorphic" to describe this alga's development cycle. Male or female gametophytes produce gametes and are at the origin of sexual reproduction in this species.

The cultivation of *L. japonica* is based on the controlled production of sporophytes and gametophytes (see Figure 2.40). The cultivation process begins with the selection of sporophyte progenitors at sea, which are then selected for the production of flagellate spores or zoospores. These mobile spores are contained in specific zones called sori, found on the distal part of the blades of mature individuals. The zoospores released from the sori are cultivated in basins located in hot greenhouses. They rapidly migrate to the bottom and lose their flagella. The spores then produce a filament and begin a germination process that leads to the formation of male or female gametophytes. The production of gametophytes, reproduction and seedling generation are carried out in this hatchery and nursery system. All greenhouse operations are carried out between July and October. During this phase, the seawater temperature is constantly adjusted via chillers. Luminosity and photoperiod are controlled by blackout systems (roofs, black or white curtains). The land phase is completed by obtaining young sporophytes suitable for transplantation to the marine rearing site. As a general rule, young sporophytes are transferred to the open sea in mid-November. This marks the start of the grow-out phase, which lasts until mid-July. This phase may require the addition of nitrates via spraying, if nitrogen levels in the water are below 100 mg.m^3. The algae are generally implanted vertically on ropes (see Figure 2.40). However, other variants, such as horizontal or mixed cultivation, are also possible.

When the water temperature reaches 21°C, the mature kelp are harvested for food or for use as broodstock in the cultivation process.

Another brown seaweed, *Undaria pinnatifida* or wakame (see Figure 2.41), is frequently produced in Asia (China, Japan, South Korea). Unlike *L. japonica*, it is only used in direct human feeding, i.e. as a sea vegetable. This Asian-grown seaweed is also produced via algoculture in

Brittany and Normandy. Its production is at the origin of the introduction of large seaweed cultivation in France and Europe. The cultivation method developed in the 1980s by Ifremer researchers (Perez 1997) is very different from that used in Asia. The Asian method is based on a number of time-consuming and labor-intensive outdoor stages. Six of these stages are detailed in Table 2.8.

The French method is based on a more limited number of steps and requires fewer people to carry it out. This method, known as the free-living method, is based on culturing sporophylls or fertile bases (see Figure 2.41) recovered from mature individuals (sporophytes) in a confined environment (six-liter culture flask). The first phase of the French method involves producing male and female gametophytes in a culture flask (see Figure 2.42). The initially clear contents of the flask rapidly darken over the course of several days, until they soon become opaque. This gametophyte mother culture is then distributed into other flasks filled with a suitable nutrient solution. At this stage, the gametophyte reserves are ready for gamete production and thus for sexual reproduction. During this phase, gametophytes are considered to be living freely in their culture medium without adhering to a support, hence the name "free-living" given to the French method.

Gametophyte proliferation is halted by applying lighting limited to 500 lux and increasing the temperature to 27°C at a rate of 0.5°C per day. Once this stage has been completed, the flask containing the gametophytes is subjected to high illumination (2,600 lux) and the temperature is lowered to 17°C at a rate of 0.5°C per day.

Two to three weeks later, gametophytes release clouds of male or female gametes. The exiguity of the environment and the stirring to which it is subjected favor the meeting of gametes and thus reproduction. This reproductive phase produces a large number of zygotes (10^{18} per flask). The fertile solution obtained is then sprayed onto collectors which are immersed vertically in basins (see Figure 2.42). Then, the seawater in these tanks is filtered and enriched with nitrogen (2.5 $mg.L^{-1}$) and phosphorus (0.64 $mg.L^{-1}$).

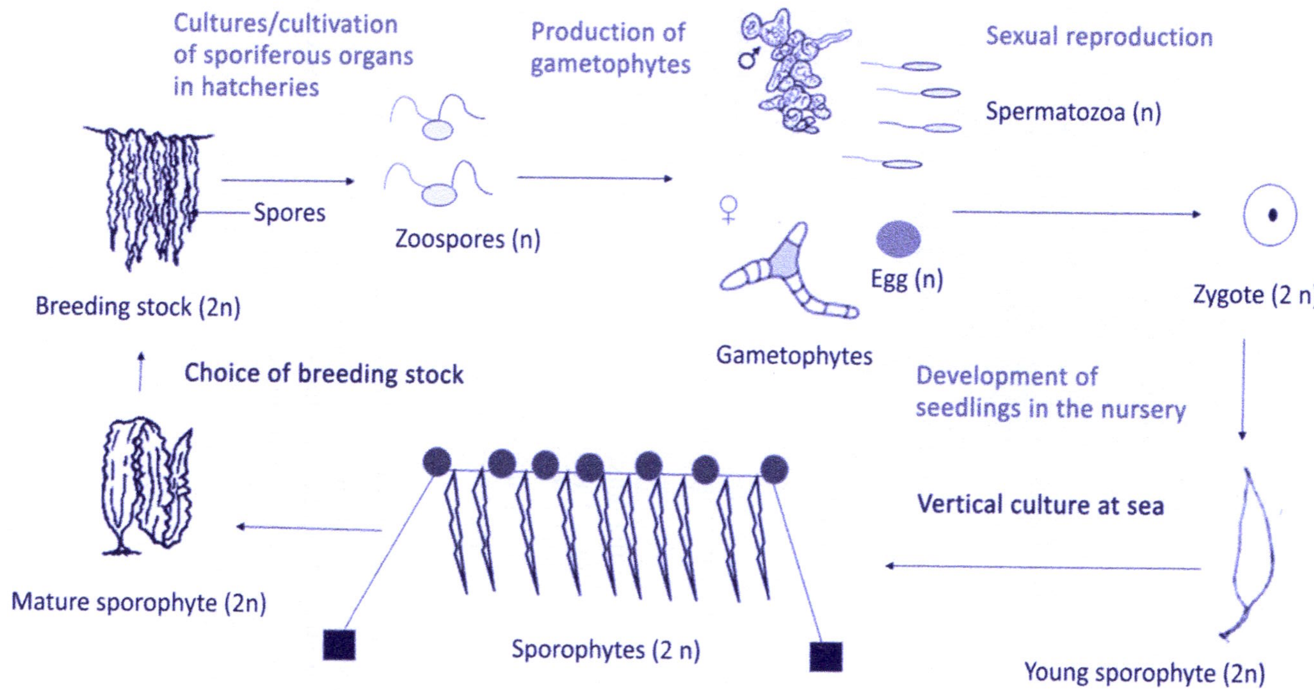

Figure 2.40. *Principle of* Laminaria japonica *cultivation (Fleurence 2022) (green: hatchery and nursery phases on land; blue: open sea phases). For a color version of this figure, see www.iste.co.uk/fleurence/aquatic.zip*

Figure 2.41. *Thallus of* Undaria pinnatifida *with its fertile base (sporophyll) on the lower part of the stipe (photo credit © Fleurence J. 2022)*

After two weeks in the tank, numerous seedlings appear on the cords of the submerged collectors. These seedlings are between 3 and 4 mm long. When the density reaches 8–10 individuals per cm, the cord can be transferred to the sea for sporophyte development. After sea culturing, the thalli attached to the string have an average size of 225 cm and an average weight of 450 g.

In addition to *Undaria pinnatifida*, other brown algae are also cultivated in Europe and North America. These are species produced for direct or indirect human use. These include the species *Alaria esculenta*, commonly known as dabberlocks, which is mainly used as a sea vegetable.

In Norway, the cultivation of *Saccharina latissima*, formerly *Laminaria saccharina*, for direct food use or alginate production (see Figure 2.43), has been expanding in recent years.

Production stage	Process	Purpose	Phase
1	Collection of sporophylls from sporophytes, air-dried at room temperature, protected from light (1–2 h)	Recovery of spore-producing organs	Nursery
2	Incubation of sporophylls in tanks containing filtered seawater water (16–18°C)	Spore release (100,000–150,000/mL)	Nursery
3	Immersion of collectors in basins	Inoculation of collectors (50–100 spores at microscopic magnification of 100)	Nursery
4	Transfer of seeded collectors to tanks containing filtered seawater	Development of seedlings to 200 µm in length	Covering
5	Open sea transfer	Horizontal cultivation on rope (line length: 8 m, space between lines: 2 m)	Growth (October–April)
6	Harvesting of thalli	Seaweed harvesting from 2 to 3 m	Harvesting for processing (80–130 kg of seaweed per line)

Table 2.8. *Main steps in the Asian method for cultivating* Undaria pinnatifida *(according to FAO (2021a))*

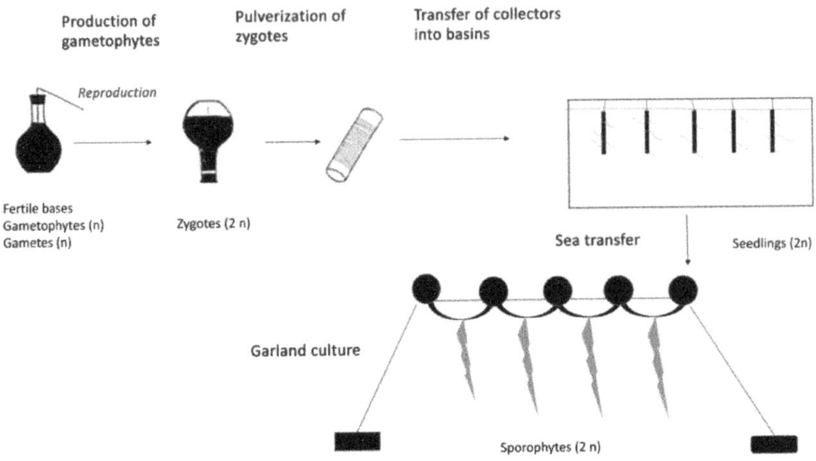

Figure 2.42. *Principle of the free-living method applied to the cultivation of the brown alga* Undaria pinnatifida *(from Perez (1997))*

Apart from brown seaweed, other species belonging to the red seaweed family are also produced by algoculture (see Figure 2.44). Some species, such as those belonging to the genus *Porphyra*, are used as sea vegetables or ingredients under the common name of nori. In particular, they are used in the preparation of maki and, as such, are well-known to most consumers, including Westerners.

Figure 2.43. *Harvesting* Laminaria saccharina *(*Saccharina latissima*) from a Norwegian cultivation site (photo credit © Stévant P. 2018)*

Species of the genus *Gracilaria* are used in two ways. They are cultivated for the production of agar, a phycocolloid widely used by the food industry, but are also consumed as vegetables in Japan under the name of ogonori.

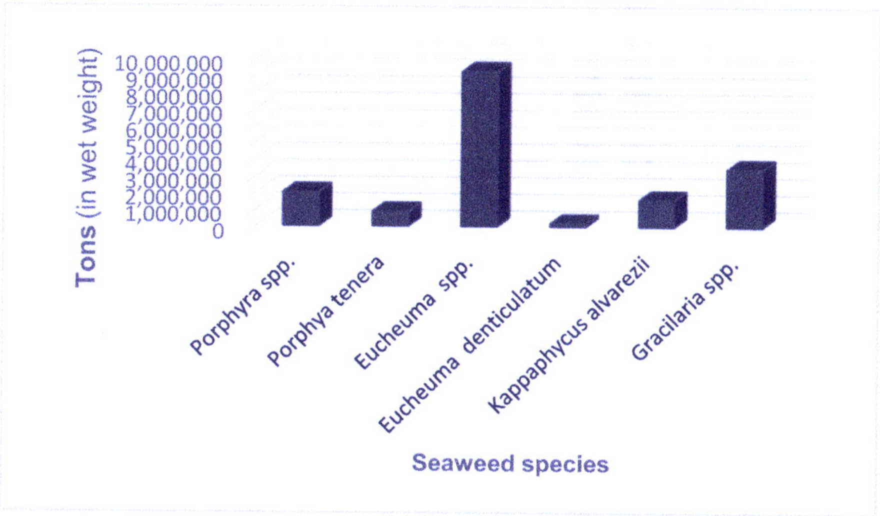

Figure 2.44. *World algal production of the main red algae used for direct human consumption (sea vegetables) or indirectly (phycocolloids) (according to FAO (2020))*

Kappaphycus alvarezii and *Eucheuma* species, meanwhile, are mainly cultivated for the extraction of carrageenan, a gelling or thickening molecule widely used as a food additive. The Philippines and Indonesia are the world's leading producers. In particular, Indonesian production has expanded significantly in recent years, rising from 4 million tons in 2010 to 11 million tons in 2018.

Among all countries, China remains the world leader in the cultivation of large algae (see Figure 2.45). Recently, Indonesia has supplanted historic Asian producers such as South Korea, Japan and the Philippines. In any case, Asia remains the main zone for the production of macroalgae by cultivation, with other countries, notably in the West, occupying a very marginal position in this field of activity.

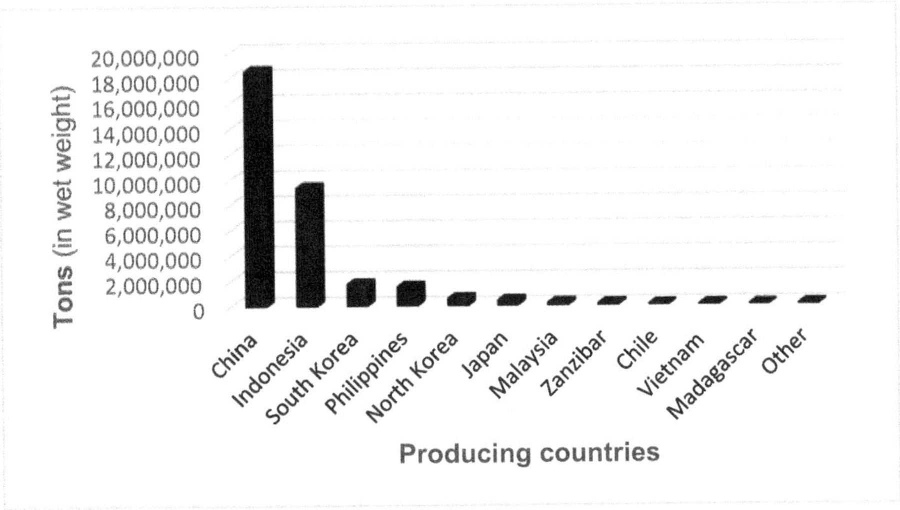

Figure 2.45. *Main seaweed-producing countries by crop (according to FAO (2020))*

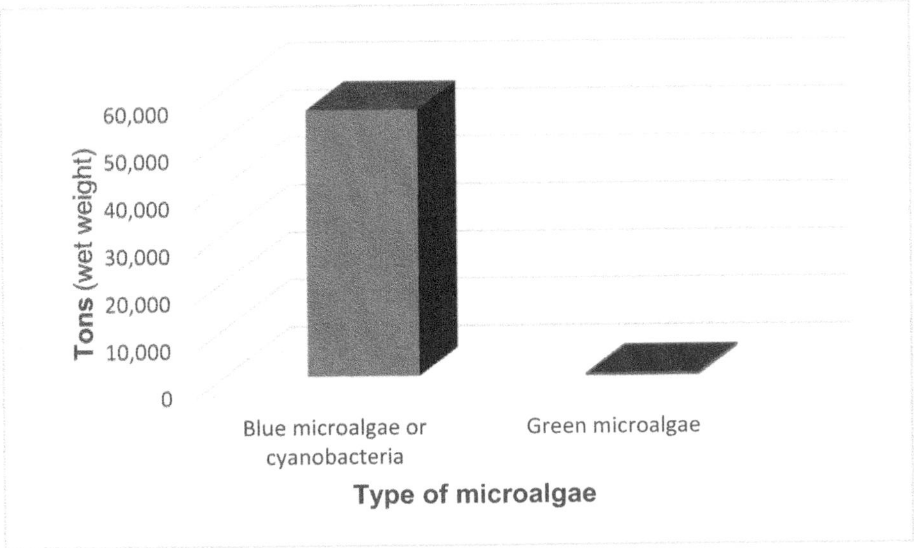

Figure 2.46. *World production of microalgae and cyanobacteria (*Spirulina *spp. or* Arthrospira *spp.) (from FAO (2021b))*

In addition to macroalgae cultivation, there is a cultivation activity dedicated to microalgae. These include spirulina, which, contrary to popular belief, does not belong to the algae group but is a prokaryotic organism belonging to the cyanobacteria family. In spite of this, spirulina is considered a microalga by the general public and official bodies. It is often referred to as blue microalgae, or simply blue algae (see section 1.2.1). In this context, the global production of spirulina (*Spirulina* spp. or *Arthrospira* spp.) in 2019 was estimated at 56,208 tons (fresh weight) (FAO 2021b). Other microalgae belonging to the green algae family (*Chlorella* spp., *Haema tococcus* spp.) accounted for just 248 tons (fresh weight) (FAO 2021b) (see Figure 2.46). Asia remains the main continent for microalgae production, including spirulina. South America, Chile in particular, is the world's second largest producer.

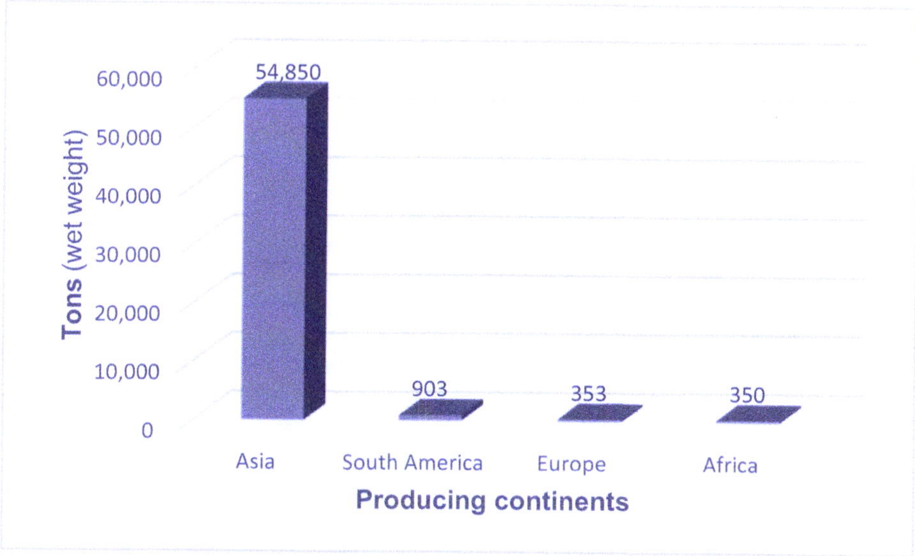

Figure 2.47. *Production of microalgae and cyanobacteria by continent (*Spirulina *spp. or* Arthrospira *spp.) (from FAO (2021b))*

Africa and Europe are far behind, producing 353 and 350 tons, respectively (see Figure 2.47). In Europe, nearly 59% of production takes place in France.

Whatever the continent or country, spirulina production is the main microalga or cyanobacterium cultivated. It accounts for almost 99% of microalgae production (FAO 2021b).

Microalgae cultivation methods vary according to their intended applications. For food applications, particularly in human or animal nutrition, the most common production method is the so-called open method. Microalgae are mass produced in natural (lakes, lagoons) or artificial (raceways) basins, in the open air or in greenhouses (see Figure 2.48) (Fleurence 2021a).

In raceway production, the culture is supplied with nitrogenous nutrients and trace elements throughout the production phase. Excess biomass is removed to avoid the shading effect caused by surface microalgae on those present in the water column. This open production method, which requires constant intervention, is known as continuous cultivation. The lighting required is generally provided by natural light, the intensity of which can be regulated by the use of blackout curtains on the greenhouse roofs.

Figure 2.48. *Production of spirulina (*Arthrospira platensis *or* Spirulina platensis*) in open raceway ponds in a greenhouse (photo credit © Westspiruline, Velly.S. 2019)*

Spirulina can also be produced in a closed system in tubular photobioreactors on a continuous basis with a permanent supply of CO_2, minerals and a controlled flow rate (see Figure 2.49). Illumination can be provided naturally (solar) and regulated by a system of blackout curtains. Artificial lighting using neon LEDs can also be used to replace natural lighting.

Figure 2.49. *Production of spirulina (*Spirulina platensis *or* Arthrospira platensis*) in a vertical tubular solar photobioreactor (photo credit © IGV Biotech 2013, Wikimedia/CC BY-SA 3.0)*

Photobioreactor production is mainly used when biomass is destined for biotechnological applications such as the production of pigments (e.g. phycobiliproteins) or high-value-added molecules (e.g. polysaccharides, enzymes, polyunsaturated fatty acids) (Fleurence 2021a).

The same cultivation processes, in open or closed systems, are also applied to the production of chlorella (*Chlorella vulgaris*), the second largest microalgal resource produced worldwide.

The production of microalgae is based almost exclusively on cultivation. This is based on a simple biotechnological approach: the production of a previously introduced strain of microalgae in a limited natural or artificial space (raceway-type ponds). In artificial ponds, water flow, renewal and the supply of minerals in CO_2 and light can be largely controlled. On the other hand, this type of open-air production cannot avoid the contamination of the culture by undesirable micro-organisms (bacteria, fungi, cyanobacteria, other microalgae).

Raceway cultivation of spirulina (*Spirulina platensis* or *Arthrospira platensis*) is particularly sensitive to contamination by *Spirulina minor*, another cyanobacteria. Its cultivation is also affected by potential contamination by *Chlorella* sp., an ubiquitous microalga. These two types of contamination, well-known to seaweed growers, are estimated to generate a 15–20% loss in biomass for *Spirulina platensis* grown in raceways.

However, the problem of the contamination of open cultures is not the same for all microalgae species. Indeed, the cultivation of the halophilic green alga *Dunaliella salina* largely escapes this constraint. This microalga, mainly used for the production of β-carotene, cosmetics and dietary supplements, is harvested from ponds (see Figure 2.50) or cultivated in giant ponds (1–200 ha) (Fleurence 2021a).

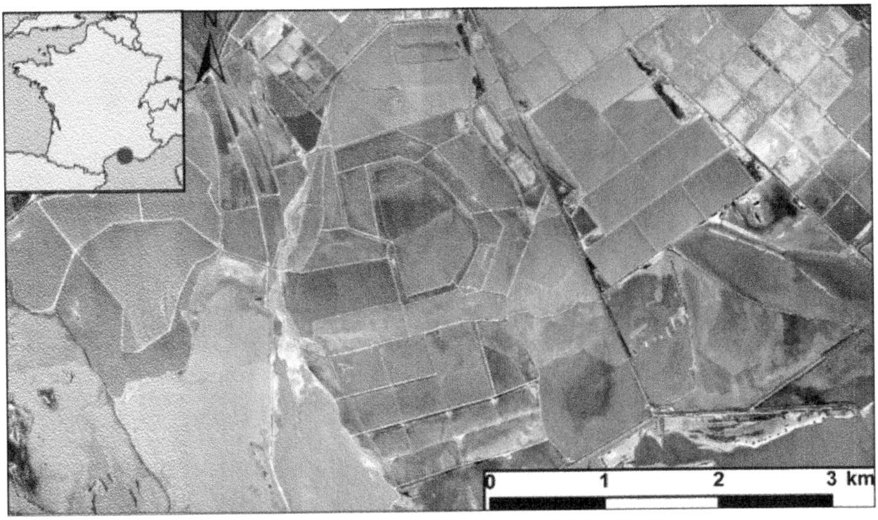

Figure 2.50. *Mass presence of the alga* Dunaliella salina *in Camargue ponds (photo credit © IGN, Brunier G. 2020 (modified))*

The high salinity of the water in the ponds limits the contamination of the *D. salina* population by other micro-organisms.

With the exception of this very special case, only closed-system production technology is likely to prevent the contamination of microalga cultures by other exogenous organisms. Despite this, obtaining axenic cultures in photobioreactors remains a delicate operation.

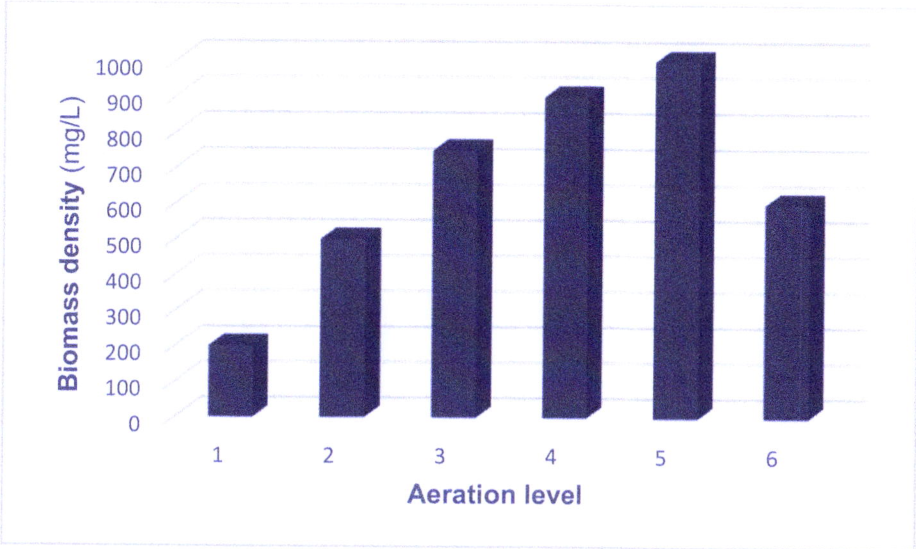

1: 0.00; 2: 0.50; 3: 0.75; 4: 1.25; 5: 2.50 min-1L-1

a)

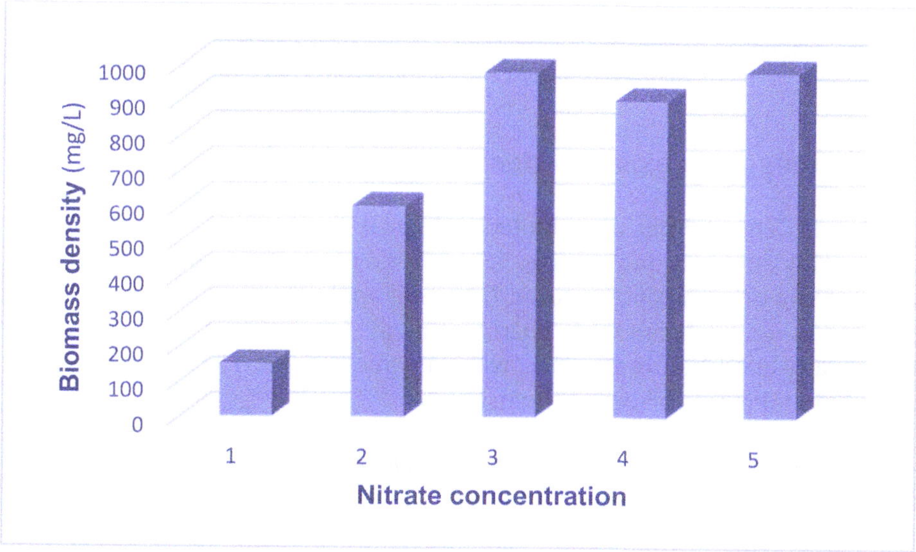

1: 0.0; 2: 0.75; 3: 1.50; 4: 2.00; 5: 3.00 mM

b)

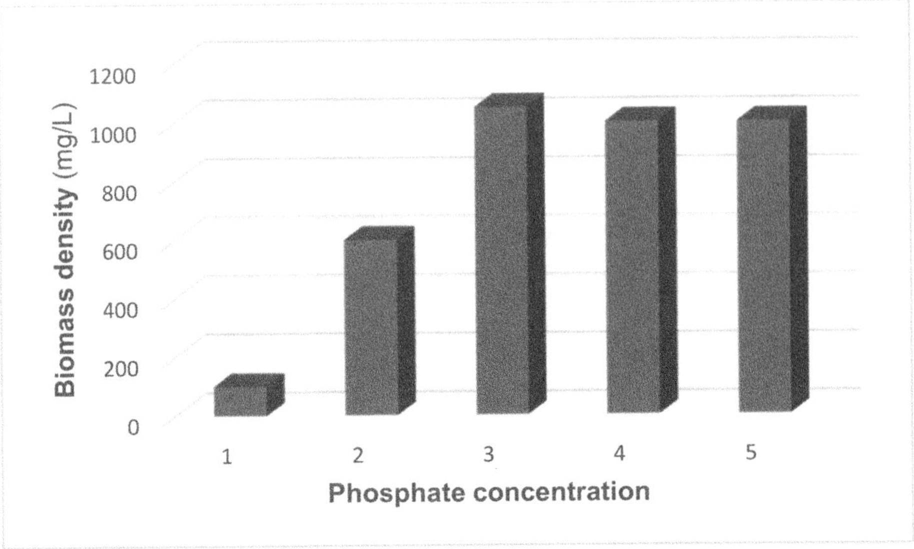

1: 0.0; 2: 0.10; 3: 0.20; 4: 0.40; 5: 0.7 mm

c)

Figure 2.51. *Effect of aeration (a), nitrate (b) and phosphate (c) supplementation on the production of transgenic gametophytes (plasmogen activator protein) of* Laminaria japonica *grown in bioreactors (from Gao et al. (2005))*

2.2.2. *Transgenic algae production*

Genetic engineering as a biotechnological approach is also applied to algae. This involves microalgae, as well as macroalgae, or at least their microscopic or gametophyte form. Expected applications mainly involve the production of molecules of pharmaceutical interest and even vaccines. Genetically modified organisms, when in microscopic form, are grown in closed photobioreactor-type systems.

The macroalga *Laminaria japonica* (*Saccharina japonica*) is a species with a heteromorphic digenetic cycle characterized by the presence of a microscopic gametophyte (see section 2.2.1). In this species, it is possible to produce transgenic gametophytes incorporating the gene coding for plasminogen-activating protein (rtPA), a factor that induces fibrinolysis. Fibrinolysis, which plays a key role in clot destruction, is an essential cardiovascular protection mechanism. The cultivation of genetically modified gametophytes and optimal conditions for the production of plasminogen activators are now well-established (Gao et al. 2005). Three criteria – aeration of the culture in the bioreactor and nitrate and phosphate supply – are decisive factors in biomass production and hence in the production of the activator protein (see Figure 2.51). Under optimal growth conditions for transgenic gametophytes, rtPA production yields 50 $\mu g.g^{-1}$ of dry cell biomass (Gao et al. 2005).

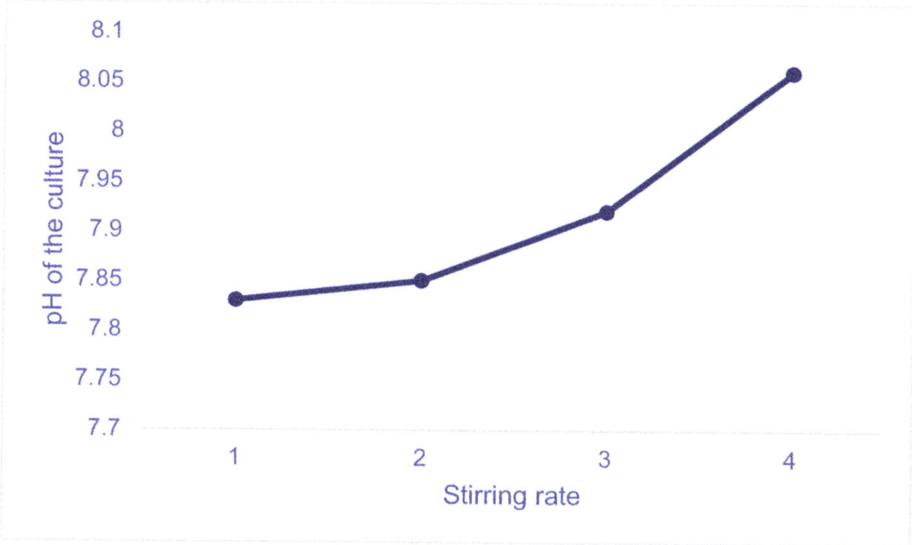

1: 100 rpm; 2: 250 rpm; 3: 150 rpm; 4: 200 rpm

Figure 2.52. *Effect of stirring rate on the pH of the gametophyte culture of* Laminaria japonica *(*Saccharina japonica*) produced at 15°C under an illumination of 138.5 $\mu Em^{-2}s^{-1}$ over 30 minutes (from Gao et al. (2006))*

This biotechnological approach has also been the subject of another medical application. The production of hepatitis B virus surface antigens (HBsAg) by transformed gametophytes of *L. japonica* with a view to the development of future vaccines remains a potential application of transgenesis employed with macroalgae (Gao et al. 2006). Temperature, culture stirring, illumination and culture density are determining factors for the optimized production of gametophytes transformed via the viral gene.

These factors depend in particular on the pH variation of the culture during the production process. pH remains an essential factor in the viability and growth of gametophytes (seawater pH: 8.05–8.10). Measuring the pH and its evolution is therefore a quick way to monitor and ensure good growth conditions for transgenic gametophytes in the culture. The optimum temperature for this type of culture is established at 15°C. At this temperature, the stirring speed of the culture has a very marked influence on the pH of the medium. After 30 minutes of stirring, a speed of only 30 rotations per minute (rpm) allows a pH close to the average pH of seawater to be obtained (see Figure 2.52).

The effect of illumination on culture pH depends not only on its intensity but also on the initial cell density (see Figure 2.53). For a cell biomass density of 135.6 $mg.L^{-1}$, the effect on culture pH is most pronounced at an illumination level of 138.5 $\mu Em^{-2}.s^{-1}$ (see Figure 2.53(a)). For a higher biomass density, this level of illumination produces a lesser effect on the pH increase (see Figure 2.53(b)).

The cultivation of transgenic gametophytes in photobioreactors is therefore possible in the same way as for microalgae. As with microalgae, optimizing their production in a closed, controlled system requires in-depth engineering. This type of approach can therefore only be envisaged in the context of the mass production of high-added-value molecules.

Transmission of the transgene from the transformed gametophyte to the adult form or sporophyte is also possible. As with many organisms, this is a case of transgenesis inherited either by fertilization (see Figure 2.54) or possibly by parthenogenesis (see Figure 2.55).

In *L. japonica*, this procedure shows that the transgenes present in gametophytes can be transmitted and expressed in sporophytes. This was verified in particular for the gene encoding HBsAg. The transgenic sporophytes obtained had a level of HBsAg gene expression comparable to that recorded with transgenic tobacco (Qin et al. 2005). The average antigen content produced was 0.561 µg/mg of soluble protein. Under certain conditions not detailed by the authors, the maximum HBsAg concentration could reach 2.497 µg/mg of soluble protein.

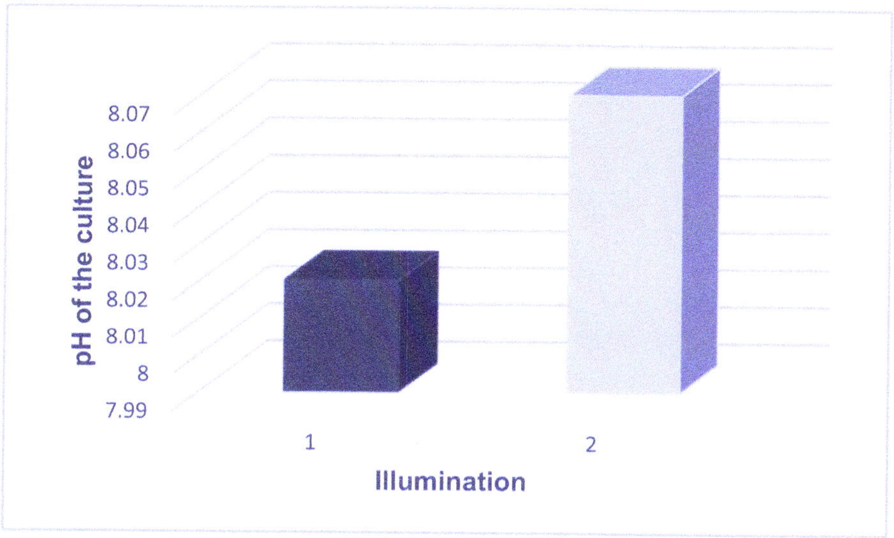

1: 227.6 µEm^{-2}s^{-1}; 2: 138.5 µEm^{-2}s^{-1}. Cell density: 135.6 mg.L^{-1}

a)

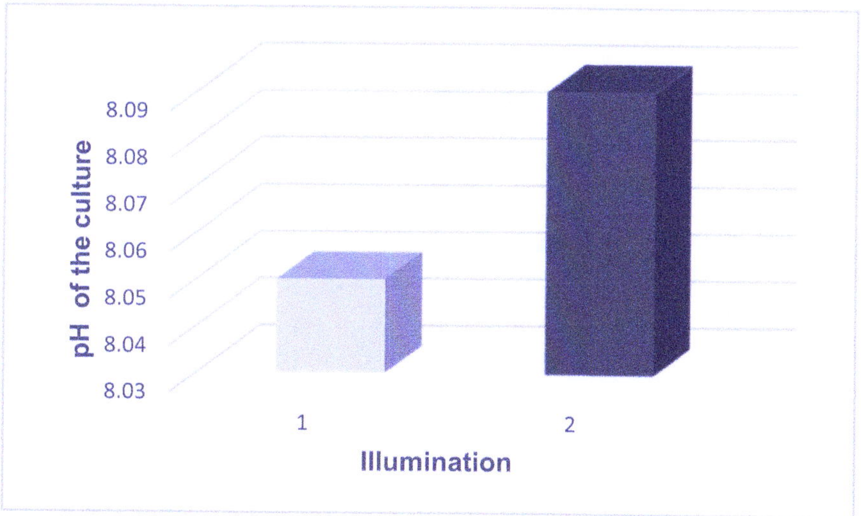

1: 146.8 µEm^{-2}s^{-1}; 2: 273.4 µEm^{-2}s^{-1}. Cell density: 273.6 mg.L^{-1}

b)

Figure 2.53. *Effect of illumination on pH in cultures of transgenic gametophytes of* Laminaria japonica *(*Saccharina japonica*) in relation to cell density (from Gao et al. (2006))*

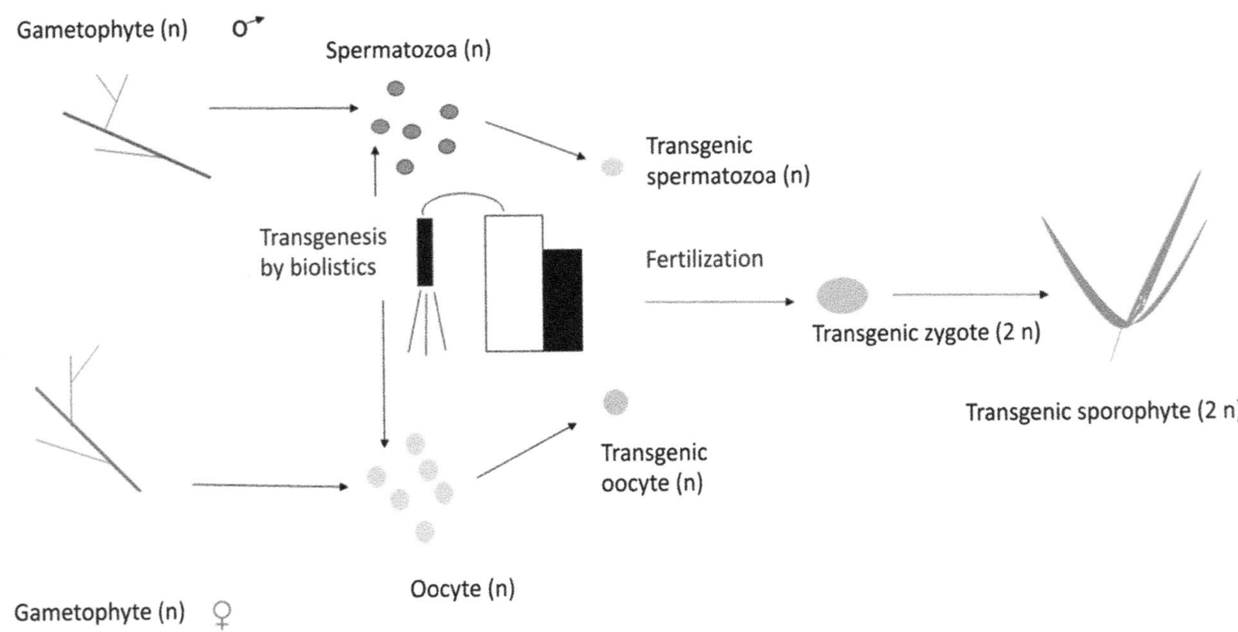

Figure 2.54. *How to obtain transgenic sporophytes of* Laminaria japonica *(*Saccharina japonica*) after the application of the biolistics technique for gene transfer in algal germ cells and fertilization (Fleurence (2022) after Qin et al. (2005))*

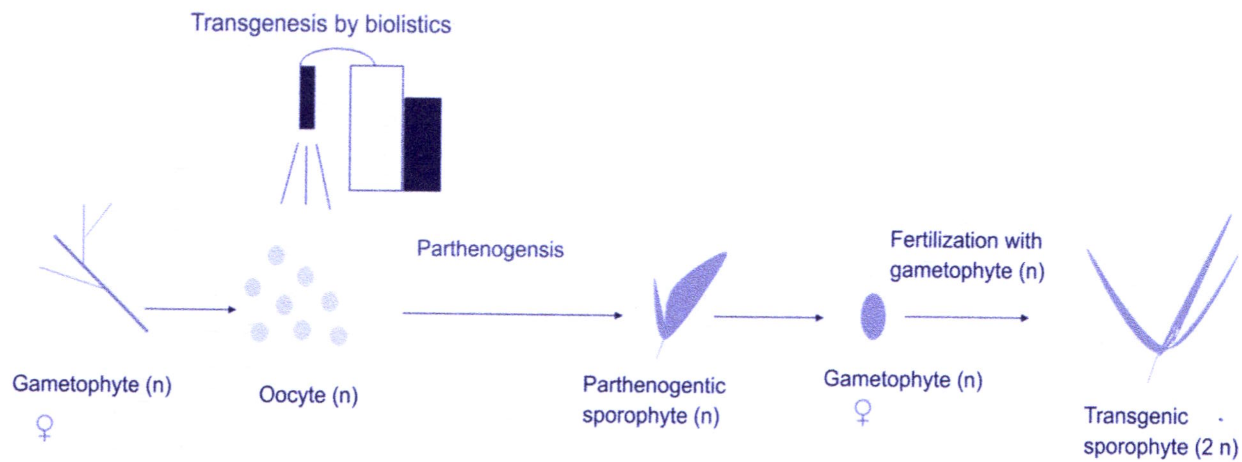

Figure 2.55. *Method of obtaining transgenic sporophytes of* Laminaria japonica *(*Saccharina japonica*) following the application of the biolistics technique for gene transfer into algal germ cells and parthenogenesis (Fleurence (2022) after Qin et al. (2005))*

The stability of the transgene in the genome, whether nuclear or chloroplastic, is a decisive element for its expression and for the establishment of transgenesis heritable by descendants.

Among macroalgae, the main transformed algae belong to the genera *Laminaria*, *Undaria* (brown algae), *Ulva* (green algae), *Porphyra* and *Gracilaria* (red algae). The species *L. japonica* and *U. pinnatifida* are characterized by stable genetic transformation (see Table 2.9).

Of the other species, only the red alga *P. yezoensis*, known by the name nori, shows genetic stability to processing (see Table 2.9) (Hallmann 2007).

Species	Group	Quality of genetic transformation
Laminaria japonica	Brown seaweed	Stable
Undaria pinnatifida	Brown seaweed	Stable
Ulva lactuca	Green seaweed	Transient
Porphyra yezoensis	Red seaweed	Stable
Porphyra miniata	Red seaweed	Transient
Gracilaria changii	Red seaweed	Transient
Kappaphycus alvarezii	Red seaweed	Transient

Table 2.9. *Stability of genetic transformation for algal species that have undergone a transgenesis procedure (according to Hallman (2007))*

The biotechnological processes used to transfer exogenous genes into macroalgae are identical to those used for terrestrial plants. In addition to bombardment with particles carrying the gene to be transferred, *Agrobacterium tumefasciens* transformation and electroporation are also used. Electroporation has been successfully used to transform the red alga *Porphyra leucosticta* (Lin et al. 2004). Transformation via *Agrobacterium tumefaciens* has proved to be an effective process for introducing the kappa-carrageenase gene from *Pseudoalteromonas* into the carrageenophyte *Kappaphycus alvarezii* (Rajamuddin et al. 2016). Unfortunately, the survival rate of algae transformed in this way and their ability to produce a new bud are relatively low (36%).

Despite this, this transformation methodology has recently been tested on *K. alvarezii* with the aim of transferring to developing algal cells (callus) the G alpha-gene derived from terrestrial plants, which confers a certain tolerance to biotic and abiotic stresses (Sulistiani et al. 2019).

Transgenesis is mainly based on the transformation of genetically modified gametophytes cultivated in closed photobioreactor-type systems. Nevertheless, the cultivation of transgenic sporophytes is possible, and their potential dissemination in an open environment such as the oceans poses an ecological problem. As with transgenic salmon, the promoters of this concept suggest cultivating transgenic algae in confined environments, such as land-based ponds.

The main applications for transgenic macroalgae concern the production of molecules for human health (vaccines, proteins). In this respect, transformed algae appear to be an additional biological factory, on a par with transgenic terrestrial plants. Despite this, other applications linked to the agri-food industry are envisaged via this genetic engineering approach (see Table 2.10).

These applications are those usually achieved with macroalgae in the food domain (production of phycocolloids and sea vegetables). As in the case of AquAdvantage salmon, the possibility of promoting the development of transgenic edible algae therefore remains a possibility. However, the production by cultivation of non-genetically modified algae meets the current needs of the agri-food market and the low costs associated with this type of market. This observation could rule out this possibility for the moment.

Processed species	Botanical group	Potential applications
Laminaria spp.	Brown seaweed	Alginate production Production of food biomass (kombu)
Undaria spp.	Brown seaweed	Food biomass production (wakame)
Gracilaria spp.	Red seaweed	Production of food biomass (ogonori) Agar production
Porphyra spp.	Red seaweed	Production of food biomass (nori) Cosmetic active ingredients
Kappaphycus spp.	Red seaweed	Carrageenan production
Ulva spp.	Green seaweed	Food biomass production (aonori)

Table 2.10. *Some examples of potential applications outside the human health sector of genetically modified algae (from Hallman (2007))*

Unicellular algae and cyanobacteria are organisms that naturally produce molecules of varied interest, such as pigments, enzymes, exopolysaccharides and fatty acids (Fleurence 2021a). The ease of cultivating them in controlled environments makes these microalgae into veritable cell factories. These microalgae can be genetically modified and grown in closed systems such as photobioreactors. There are several types of photobioreactor that can be used for closed-system cultivation of microalgae, whether genetically modified or not. These include horizontal tubular, vertical tubular and planar honeycomb photobioreactors (Fleurence 2021a). In theory, these controlled, confined systems limit the ecological risk.

As with macroalgae, the main method of transgenesis used remains biolistics. Electroporation or the use of cellular abrasives such as silicon carbide trichites is more secondary. Unlike macroalgae, the production of transgenic microalgae involves a large number of species and the genome targeted is not only the nuclear genome. This is the case for the green microalga *Chlamydomonas reinhardtii*, in which transgenesis can be successfully carried out on the nuclear, mitochondrial and chloro-plastic genomes. This species, like chlorella and spirulina, forms part of many fundamental (genomics) and development research projects (see Table 2.11).

In terms of fundamental research, certain species are included in genome projects, which aim to characterize the complete genetic maps of these organisms.

Potential value-added uses include applications relating to the production of recombinant proteins useful for vaccine development or even of molecules of interest such as pigments or lipids and derivatives that are sources of biofuel (see Table 2.11).

The production of non-genetically modified aquatic animals or algae based on aquaculture can be considered a biotechnological approach. Indeed, such production technologies are applied to living organisms and, as such, they meet the simplest definition of biotechnology.

The production of these organisms following genetic engineering transformation is also an example of qualified biotechnology, as far as blue revolution fish are concerned.

Beyond these concepts relating to biomass production, the transformation of these resources by enzymatic engineering is also an integral part of blue biotechnologies.

Species	Genome project	Valorizations
Chlamydomonas reinhardtii (green algae)	x	Production of transgenic biomass for animal feed Recombinant protein production Environmental quality monitoring Bioremediation
Dunaliella salina (green algae)	x	Carotenoid production Health food production Food supplement production Cosmetics production
Chlorella vulgaris (green algae)	x	Biomass production for health food Food supplement production Cosmetics production
Haematococcus pluvialis (green algae)	-	Carotenoid production
Nannochloropsis salina (green algae)	-	Biofuel production
Spirulina platensis (*Arthrospira platensis*) (cyanobacteria)	X	Production of phycocyanin and phycoerythrin Biomass production for health food Drug production Cosmetics production

Table 2.11. *Examples of proven or potential applications of transgenic microalgae (from Hallman (2007) and Fleurence (2021a))*

3

Biotechnological Processing Methods

3.1. Enzymatic engineering (enzymatic hydrolysis)

Enzymatic engineering covers all enzymatic processes used to transform the physico-chemical, organoleptic and nutritional characteristics of biomass intended for human or animal consumption. These processes can also be used to improve the extraction of molecules by a less denaturing route than the conventional procedure involving mechanical grinding or solvent extraction. In the latter case, the term blue chemistry is sometimes used to refer to this type of extraction.

Enzymatic engineering, as defined above, is based on the use of enzymes with hydrolytic activity. These are mainly polysaccharidases, glucosidases, proteases or lipases.

Enzymatic engineering is a biotechnological process that, when applied to the transformation of aquatic resources, whether of animal or plant origin (algae), is part of the blue biotechnology repertoire.

3.1.1. *Enzymatic engineering applied to fish matrices*

Fish-based sauces are commonly used in Asia. They are obtained by autolysis of fish flesh in a saline environment, followed by fermentation. A period of 18 months is generally needed to obtain a dark brown sauce rich in flavor-enhancing amino acids. Autolysis, the first step in the process, is carried out by endogenous proteases, and the fermentation carried out by micro-organisms also involves endogenous bacterial proteases. The addition of exogenous proteases such as trypsin and chymotrypsin (0.3% w/w) accelerates the sauce's fermentation process, which is made from a herring base (Chaveesuk et al. 1994). The fermentation time is in this case reduced by two months compared to the traditional process and without any impact on the quality of the final product.

The addition of exogenous proteases is not limited to accelerating the fermentation process for fish sauces. In particular, it is applied to the production of protein hydrolysates. Fish protein hydrolysates are considered to be high-value-added products, particularly in animal feed. They are generally obtained from protein concentrates, which are co-products of the fish canning industry, particularly tuna canning. These hydrolysates are valued as a source of peptides, free amino acids and flavors. They are obtained by proteolysis via a relatively simple process (see Figure 3.1).

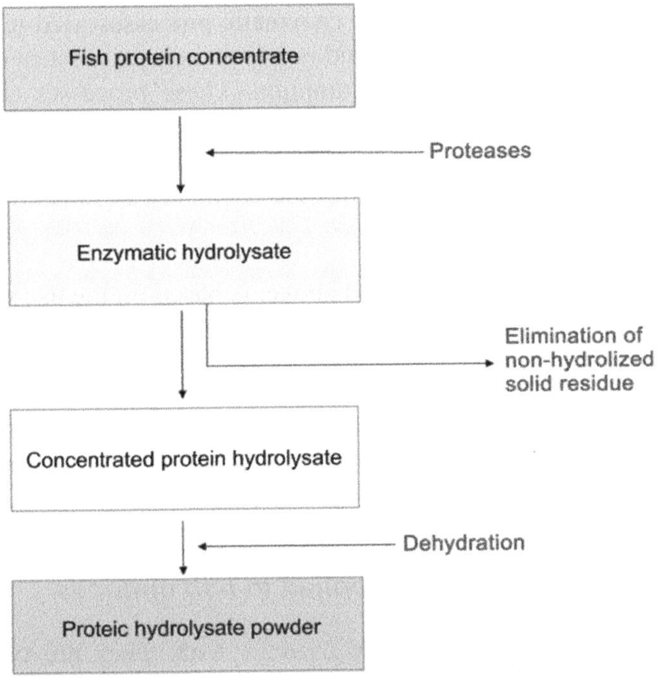

Figure 3.1. *Simplified process for obtaining a concentrated powder from fish protein hydrolysate by proteolysis (from Nilsang et al. (2005)). For a color version of this figure, see www.iste.co.uk/fleurence/aquatic.zip*

Enzymatic hydrolysis is carried out using enzyme preparations with endo- and exo-peptidase activity. Frequently used preparations are Flavourzyme, well-known for flavor production, and Kojizyme, which produces bitter-tasting aromas. These industrial proteases have many other applications, as is described later.

The hydrolysis of protein concentrates by these proteases is achieved after 6 hours (see Figure 3.2). Under the set operating conditions (40 enzyme activity units/g concentrate, 50°C), Kojizyme is more effective than Flavourzyme in terms of hydrolysis (Nilsang et al. 2005).

These protein hydrolysates are characterized by the presence of liberated amino acids, some of which, like tryptophan, are listed as essential in human or animal consumption. The use of Kojizyme significantly improves the production of free tryptophan, an aromatic amino acid involved in the development of bitter taste. In contrast, the use of Flavourzyme does not increase tryptophan release during hydrolysis time (see Figure 3.3).

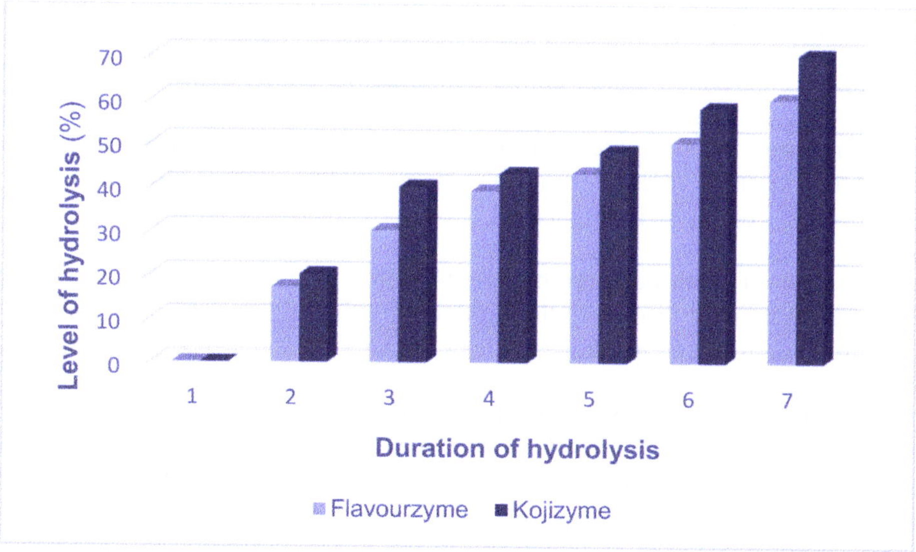

1: 0 h; 2: 1 h; 3: 2 h; 4: 3 h; 5: 4 h; 6: 5 h; 7: 6 h at 50°C

Figure 3.2. *Degree of hydrolysis of a tuna protein concentrate by Flavourzyme and Kojizyme in relation to hydrolysis time (from Nilsang et al. (2005)). For a color version of this figure, see www.iste.co.uk/fleurence/aquatic.zip*

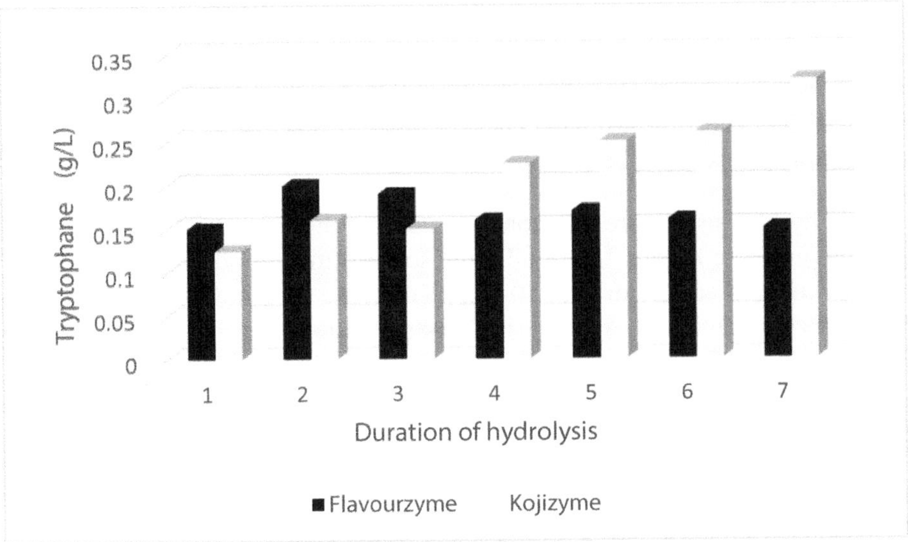

1: 0 h; 2: 1 h; 3: 2 h; 4: 3 h; 5: 4 h; 6: 5 h; 7: 6 h at 50°C

Figure 3.3. *Release of tryptophan during hydrolysis by the enzymatic preparations Flavourzyme and Kojizyme (from Nilsang et al. (2005))*

Depending on the intended application, particularly with regard to aromatic bitterness, these results provide an important indication as to the choice of process to be implemented.

The use of enzymatic engineering in fish biological matrices is an approach that is not limited to the production of peptides and amino acids of nutritional or aromatic interest.

This approach is applied to the valorization of other compounds of alimentary interest present in halieutic or aquacultural resources. It can be used to extract lipids and fatty acids from fish, avoiding the oxidation processes generated by the use of high temperatures or solvents, as is the case in industry (Dumay et al. 2004).

In cod, the lipid fraction rich in fatty acids can be extracted using a process involving the proteolysis of various by-products with little or no added value (skin, roes, liver, viscera, head, etc.).

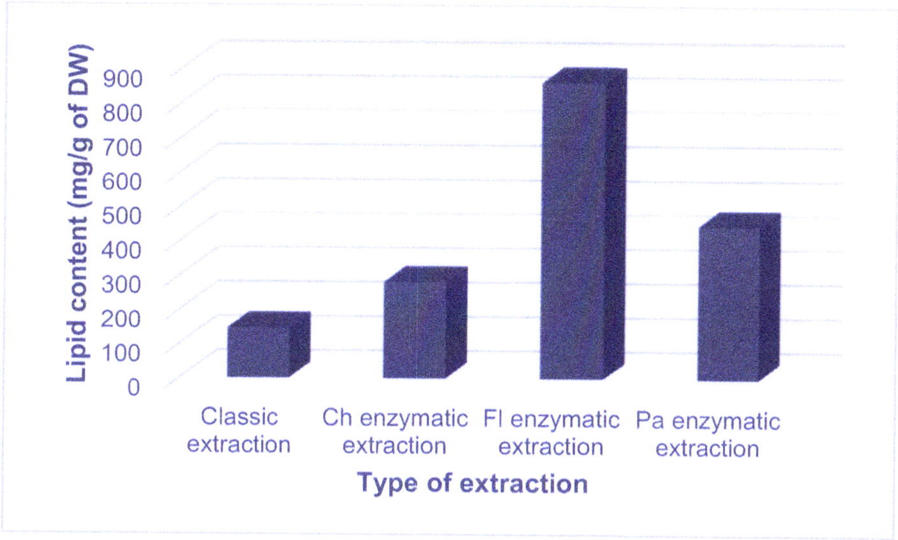

Ch: Chymotrypsin; Fl: Flavourzyme Pa: Papain

Figure 3.4. *Comparison of different types of extraction (conventional or enzymatic) of the total lipid fraction from cod (*Gadus morhua*) viscera (from Dumay et al. (2004))*

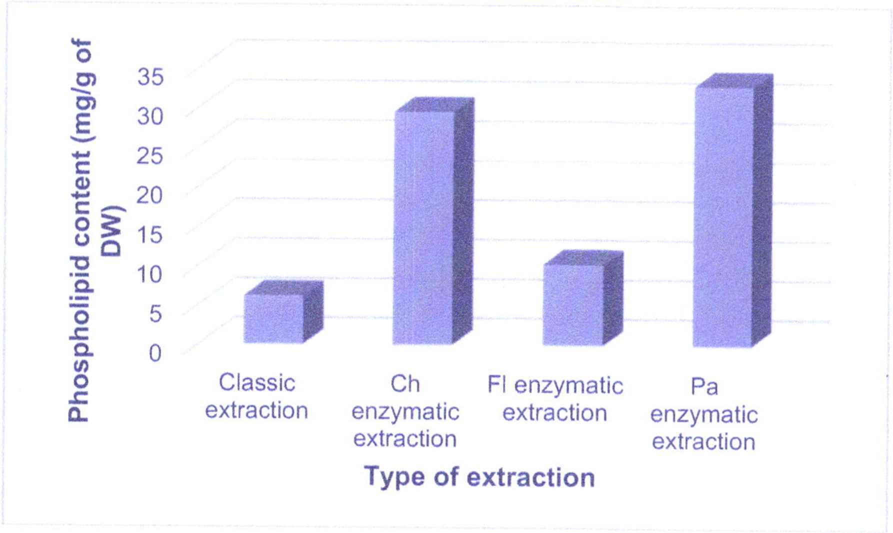

Ch: Chymotrypsin; Fl: Flavourzyme Pa: Papain

Figure 3.5. *Comparison of different types of extraction (conventional or enzymatic) of the phospholipid fraction of cod (*Gadus morhua*) viscera (from Dumay et al. (2004))*

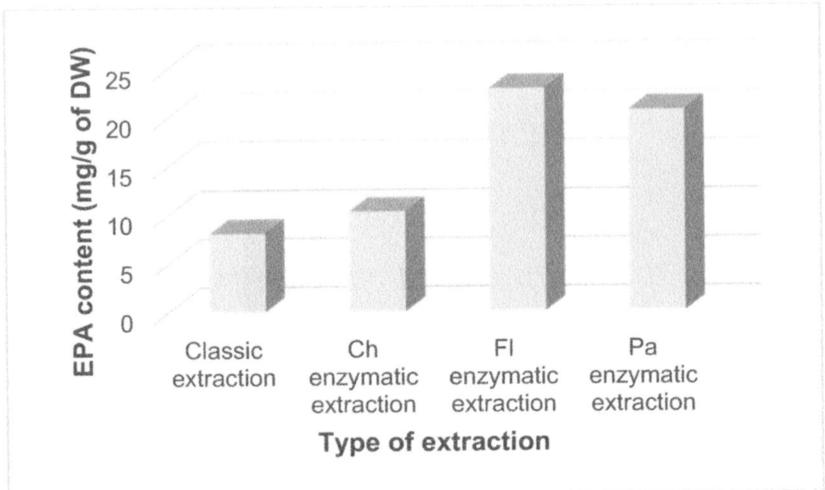

Ch: Chymotrypsin; Fl: Flavourzyme Pa: Papain

Figure 3.6. *Comparison of different types of extraction (conventional or enzymatic) of the eicosapentaenoic acid (EPA) fraction of cod* (Gadus morhua) *viscera (from Dumay et al. (2004))*

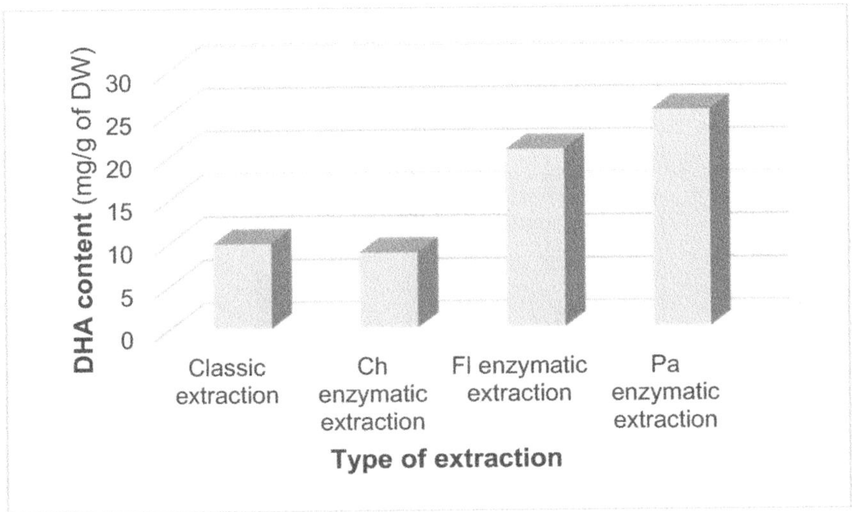

Ch: Chymotrypsin; Fl: Flavourzyme Pa: Papain

Figure 3.7. *Comparison of different types of extraction (conventional or enzymatic) of the docosahexaenoic acid (DHA) fraction of cod* (Gadus morhua) *viscera (from Dumay et al. (2004))*

Fish viscera are a potentially valuable source of phospholipids and polyunsaturated fatty acids (EPA, DHA). Treatment of these co-products by enzymatic hydrolysis significantly improves extraction yields compared to conventional processes using organic solvents (chloroform/methanol) (see Figures 3.4–3.7). The use of the industrial enzyme preparation Flavourzyme increases EPA extraction yield by 121% (see Figure 3.6) and DHA extraction yield by 136% (see Figure 3.7). However, the efficiency of the enzymatic process compared with the conventional process can differ depending on the nature of the co-product being treated. Enzymatic hydrolysis applied to fish head processing proves less efficient than the conventional process, both for the extraction of total lipids and of fatty acids such as EPA or DHA (see Table 3.1).

Processes	**Total lipids (mg/g DW)**	**Phospholipids (mg/g DW)**	**EPA (mg/g DW)**	**DHA (mg/g DW)**
Classic	53.14	13.12	3.05	3.39
Enzymatic (Flavourzyme)	50.71	7.80	1.67	1.60

Table 3.1. *Comparison between the conventional extraction process and the enzymatic hydrolysis process (Flavourzyme) with regard to the recovery yield of lipids and derivatives as applied to fish heads (from Dumay et al. (2004))*

Enzymatic hydrolysis of fishing co-products is a useful approach for extracting compounds of nutritional interest, such as lipids and polyunsaturated fatty acids. It is less denaturing and more respectful of the environment, as it does not use organic solvents. Its efficiency is generally higher than that of conventional processes. However, the latter is dependent on the organs treated and may in some cases be less effective than a conventional procedure.

This reminds us that enzymatic hydrolysis remains a process that needs to be refined in terms of the choice of enzymes to be used, and also of the matrix to be used.

In addition to food applications, enzymatic engineering can also be used to produce therapeutically active biopeptides from fish flesh. This approach forms part of the development of functional and nutraceutical foods for human consumption (Ryan et al. 2011).

Certain angiotensin-converting peptide inhibitors have the effect of lowering blood pressure and are therefore considered effective agents in the fight against hypertension. The presence of such inhibitors has been demonstrated in many species of fish, including tuna, bonito, salmon, sardines and catfish.

Proteolytic treatment of the flesh of these fish also produces hypotensive peptides of widely varying biochemical natures and inhibitory efficacy (see Table 3.2).

Species of fish	Peptides (sequence)	Matrix	Proteolytic enzyme	Inhibition concentration 50% (μM)
Bonito	IKPLNY	Muscle	Thermolysin	43
Salmon	WA	Muscle	Thermolysin	277.3
	VW			2.5
	WM			96.6
	MW			9.9
	IW			4.7
	LW			17.4
Sardines	MF	Muscle	Alcalase	44.7
	RY			51.0
	MY			193.0
Tuna	PTHIKWGD	Muscle	Pepsin	n.d.
	GDLGKTTTVSNWSPPKYKDTP	Protein fraction		11.3
Alaskan pollock	GPL	Skin	Alcalase + pronase + collagenase	2.65
	FGASTRGA	Protein fraction	Pepsin	14.7

Table 3.2. *Some examples of hypotensive peptides produced by enzymatic hydrolysis of fish flesh or skin (from Ryan et al. (2011))*

Salmon flesh subjected to proteolysis via thermolysin releases many angiotensin-converting enzyme inhibitors, including the VW peptide, well-known for its hypotensive properties. A concentration of 2.5 μM of VW peptide inhibits 50% of angiotensin-converting enzyme (ACE) activity (IC_{50}), revealing its strong inhibitory activity.

However, the production of hypotensive peptides by proteolysis is not confined to the processing of fish flesh, and it is also used to noble effect in human nutrition. The hydrolysis of skin, and collagen in particular, by an enzymatic cocktail of alcalase, pronase and collagenase yields the GPL peptide, which also has very high conversion enzyme inhibitory activity (2.65 μM) (see Table 3.2).

This route, like the previous one, can be used for the valorization of co-products from fishing and aquaculture.

The production of angiotensin-converting enzyme-inhibiting peptides by proteolysis is also possible from processed fish products that are quite different from the raw resource or co-products.

Enzymatic digestion of a traditional Japanese product, katsuobushi, made from dehydrated bonito meat, resulted in the production of eight peptides with enzyme-converting activity (Fujita and Yoshikawa 1999).

Figure 3.8. *Katsuobushi (photo credit © Midori S. 2006, Wikimedia/CC BY-SA 3.0)*

These peptides result from the proteolysis of the initial product by the action of the thermolysin. Among them is the peptide LKPNM, with an IC_{50} of 2.4 μM. This peptide can in turn be digested to give another peptide, LKP, with an IC_{50} of 0.32 μM. These two orally administrable peptides act on systolic blood pressure by inducing a negative variation in the latter (see Figures 3.9 and 3.10). Depending on the peptide administered (LKPNM or LKP), this fall in blood pressure reaches the optimal level after between 2 and 4 hours. The LKP peptide, however, appears to have a more rapid effect, producing the optimal result as early as 2 hours after administration (see Figure 3.10). Incidentally, digested katsuobushi is currently considered a health food. Marketed under the name Katsuobushi oligopeptide, it has been awarded the health food "birth certificate" by the Japanese Ministry of Health and Welfare.

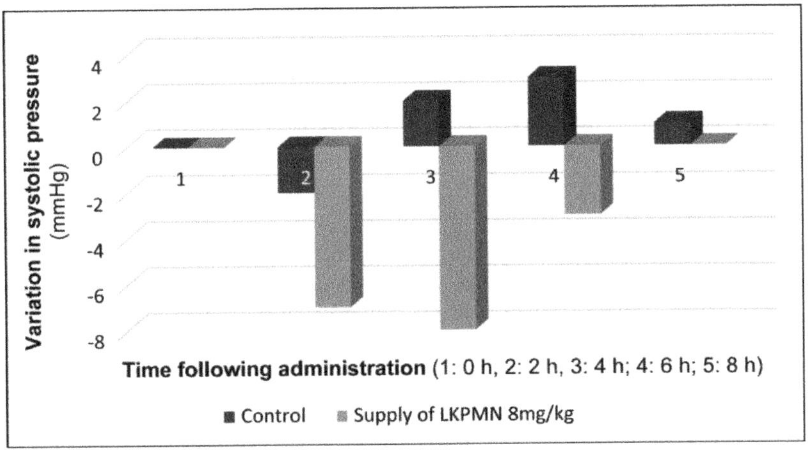

Figure 3.9. *Variation in systolic blood pressure in the presence and absence of LKPMN from enzymatic hydrolysis of katsuobushi (from Fujita and Yoshikawa (1999)). For a color version of this figure, see www.iste.co.uk/fleurence/aquatic.zip*

In addition to the use of exogenous enzymes for the transformation of fish matrices or the production of molecules of interest, certain enzymes derived from the tissues of fish or aquatic invertebrates can be used in enzymatic engineering. These enzymes are mainly proteases (see Table 3.3) (Shahidi and Kamil 2001). They are of gastric, intestinal or muscular origin. As with exogenous enzymes, they are often used for processing fishery and aquaculture co-products (see Table 3.4).

In particular, they can be used in the selective tissue degradation process, facilitating the extraction and purification of fish eggs via a gentle route. This approach is particularly useful for caviar production (see Table 3.4). Selective tissue hydrolysis is frequently used when mechanical removal is likely to damage the organ to be upgraded (e.g. cod liver). In addition to this aspect, endogenous proteases are also used for the production of fish protein hydrolysates and the release of characteristic flavors.

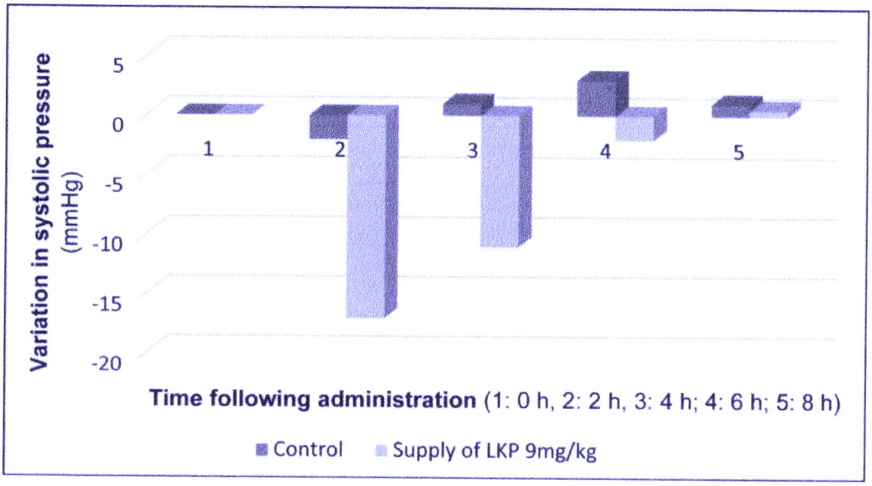

Figure 3.10. *Variation in systolic blood pressure in the presence or absence of LKP from enzymatic hydrolysis of katsuobushi (from Fujita and Yoshikawa (1999)). For a color version of this figure, see www.iste.co.uk/fleurence/aquatic.zip*

Organisms	Gastric	Intestinal	Muscular	Enzymes
Sardines	+	–	–	Pepsin
Mackerel	+	–	–	Pepsin
Cod	+	–	–	Pepsin
Salmon	+	–	–	Pepsin
Capelin	–	+	–	Trypsin
Anchovies	–	+	–	Trypsin
Rainbow trout	–	+	–	Chymotrypsin
Shrimp	–	+	–	Collagenase
Carp	–	+	+	Elastase Cathepsin A
Tilapia	–	–	+	Cathepsin B Calpains
Squid	–	–	+	Cathepsin C

Table 3.3. *Some examples of proteolytic enzymes from fish and marine invertebrate tissues and organs (from Shahidi and Kamil (2001))*

Fields of application	Examples
Selective tissue degradation	Caviar production Cod liver cleaning Removal of skin from fish and aquatic invertebrates Removal of crustacean shells
Fermentation	Production of fish sauces
Protein hydrolysate production	Protein hydrolysates and fish flavorings
Pigment extraction	Pigment recovery from shellfish waste
Other applications	Production of polyunsaturated fatty acids

Table 3.4. *Some examples of enzymatic engineering applications developed with fish matrices or aquatic invertebrates using endogenous enzymes (from Shahidi and Kamil (2001))*

In addition to fish matrices, enzymatic engineering is also applied to the transformation of other aquatic resources, such as algae, both microalgae and macroalgae.

3.1.2. *Enzymatic engineering applied to algal matrices*

Enzymatic hydrolysis of the algal matrix is a well-known biotechnological process used in a variety of contexts in both research and resource development. In research, specific hydrolases (xylanases, cellulases, carrageenases, agarases) provide a useful method for characterizing the presence or absence of certain polysaccharides in the algal cell wall.

In the field of value addition, hydrolases specific to the degradation of parietal polysaccharides are used to improve the extraction of molecules of interest (proteins, pigments, nucleic acids) or to produce bioactive oligosaccharides. The use of polysaccharidases can also be used to modify the organoleptic qualities (texture, flavor) or nutritional value of algae for human consumption. In addition to these fields of application, which concern "fine chemicals" or agri-food valorization, there is another field of application of enzymatic engineering related to the algal matrix. This field concerns the production of algal protoplasts which can be used as a biological starting stock for the production of algae by micropropagation or in vitro culture.

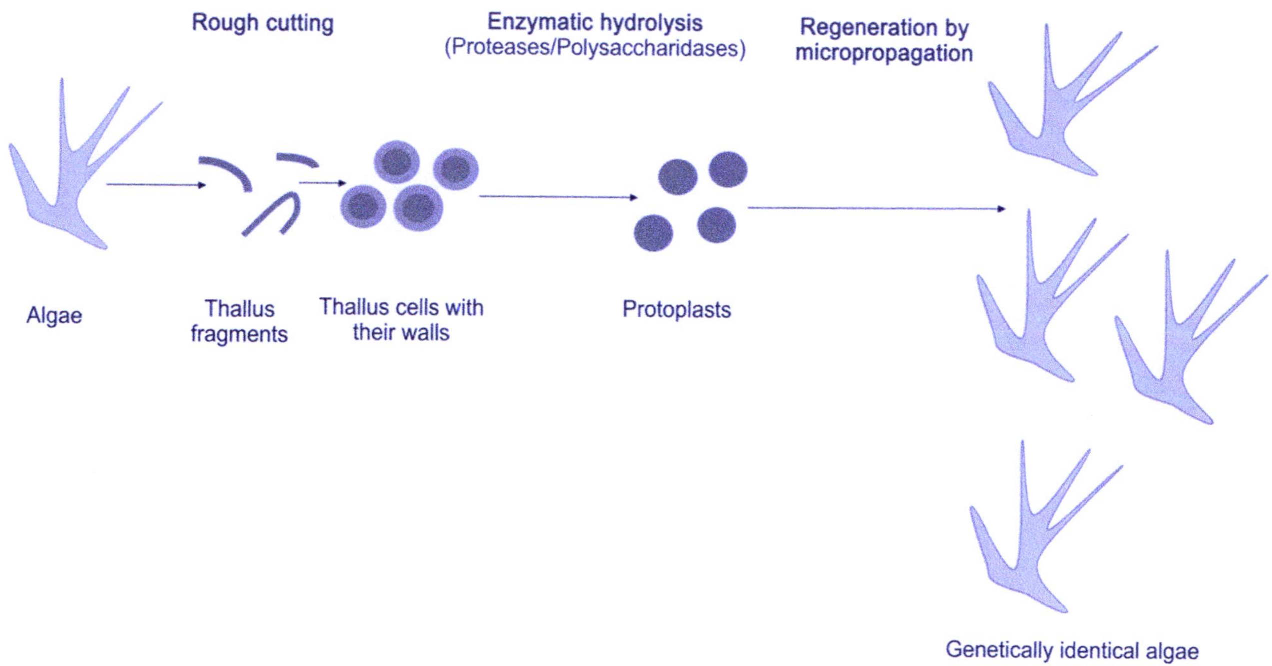

Figure 3.11. *Principle of protoplast production via enzymatic engineering as part of the development of an in vitro culture system (micropropagation) for a marine macroalga of the genus* Ulva *(Fleurence 2023)*

Algal protoplasts are cells without a cell wall. Their cultivation produces a homogeneous algal biomass of controlled quality. This biomass represents a biological resource that can be genetically improved. The regeneration of a complete organism from protoplasts is a major challenge for the production of genetically identical algae and therefore for obtaining a "standardized" biological resource (see Figure 3.11).

In green algae such as ulvae, protoplasts are generally obtained via fragmentation of the thallus, which is morphologically poorly differentiated (see Figure 3.11). They constitute an easily cultivated cellular biological resource (see Figure 3.12).

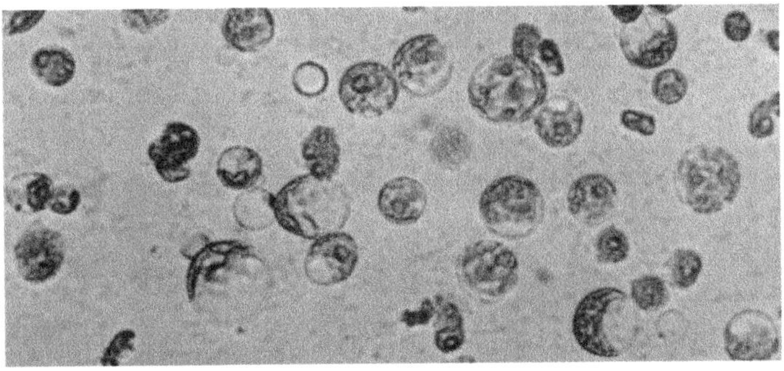

Figure 3.12. *Protoplasts of* Ulva *sp. cultivable in vitro (micropropagation) (photo credit © Rusig A.M. 2023)*

In *U. fasciata*, the production of protoplasts is used to build up a homogeneous stock of algae (Chen and Shih 2000). Protoplasts are obtained by the combinatorial use of cellulase (Onozuka R-10) (4%) and pectinases (Macerozyme R-10) (2%). The average yield of protoplast production is 3.23 x 10^7 cells per g of fresh explant. When cultured, protoplasts show a capacity to produce microthalli which, in suspension, constitute a homogeneous genetic and biological stock of *U. fasciata*.

In brown algae, such as *Laminaria japonica*, *L. digitata* or *Undaria pinnatifida*, the thallus is morphologically highly differentiated. It consists of an attachment clamp or haptera at its base; an intermediate part, the stipe, resembling a stem; and a summit part called the frond, cut into blades (see Figure 3.13). Each of these parts can be a source of protoplasts.

In *U. pinnatifida*, the optimum protoplast production yield (2–4 x 10^7 protoplasts/g fresh material) is obtained by combined hydrolysis using a commercial cellulase (1%) and an alginate lyase (4 U/mL) with a relatively short incubation time (2–4 h) (Avila-Peltroche et al. 2022). This production yield also appears to depend on the explant used (haptera, stipe, meristematic zone of the frond) (see Figure 3.14). Whole algal regeneration can be achieved after three months of culture at 20°C. The regeneration protocol enables the production of up to 30,000 sporophytes from one gram of explant.

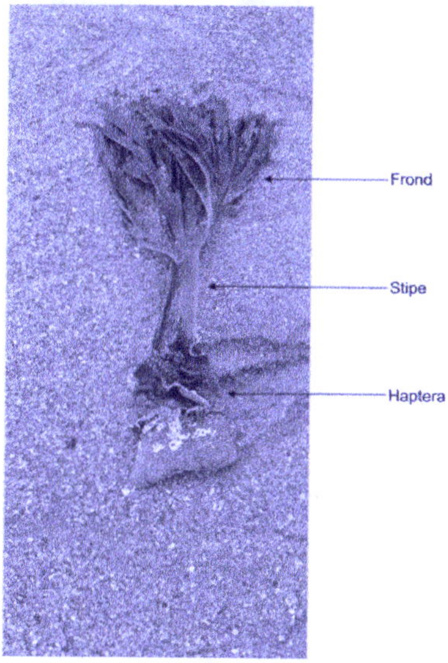

Figure 3.13. *Morphological structure of the brown alga* Laminaria digitata *(photo credit © Fleurence J. 2022)*

The proposed protocol for *U. pinnatifida* offers several advantages. The first lies in the possible reuse of the enzyme preparation used for cell wall degradation. According to Avila-Peltroche et al. (2022), the same enzymatic mixture could be used at least six times with high efficiency in terms of the viability of the protoplasts obtained and the degradation of the cell walls (see Figure 3.15). However, such reuse ultimately lowers the yield of protoplasts. Indeed, the latter is divided by almost 4.5 after the sixth use.

Another advantage of the proposed enzymatic hydrolysis process is its high efficiency compared with other processes applied to *U. pinnatifida* or even to other brown algae.

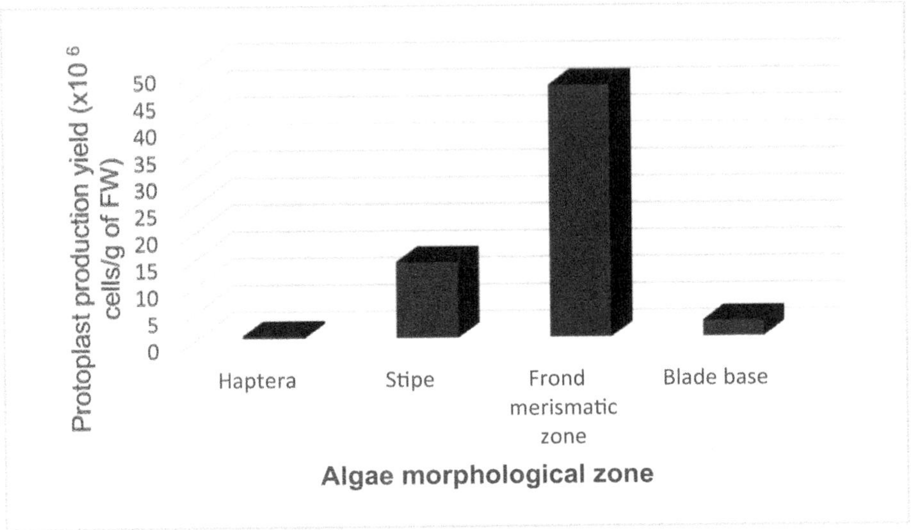

Figure 3.14. *Protoplast production yield from* Undaria pinnatifida *according to algal morphology (from Avila-Peltroche et al. (2022))*

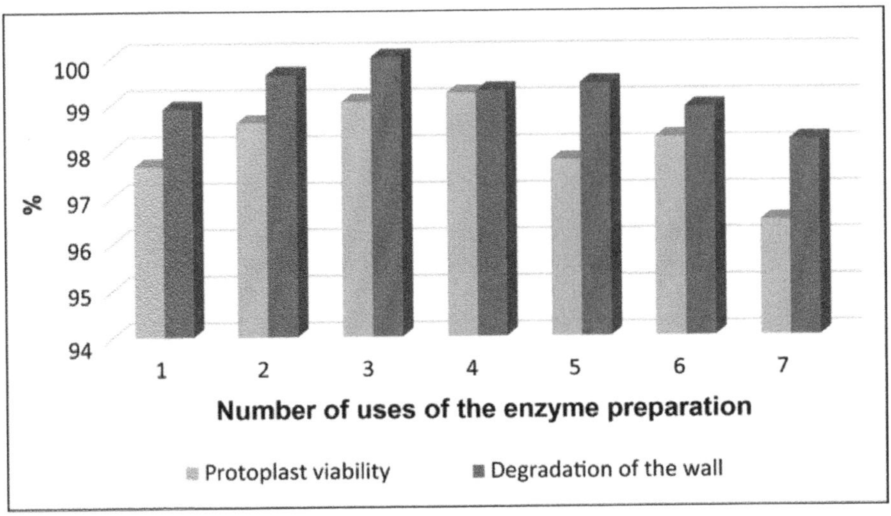

Figure 3.15. *Effect of reusing the enzyme preparation from* Undaria pinnatifida *on the viability of protoplasts and the cell wall degradation rate (from Avila-Peltroche et al. (2022))*

In particular, the combined use of cellulase and alginate lyase proves more efficient than the enzymatic hydrolysis process conventionally used, which is based on an enzymatic combination of cellulase + pectinase (cellulase Onuzuka RS + Macerozyme R-10).

The latter process, tested on various brown algae such as *Petalonia fascia* and *Sargassum duplicatum*, has also been applied to *U. pinnatifida* (Chen and Shyu 1994). It is less efficient than the alginate lyase process (see Figure 3.16). This is probably due to the absence of pectins in the wall of *U. pinnatifida*, hence the very relative effectiveness of Macerozyme.

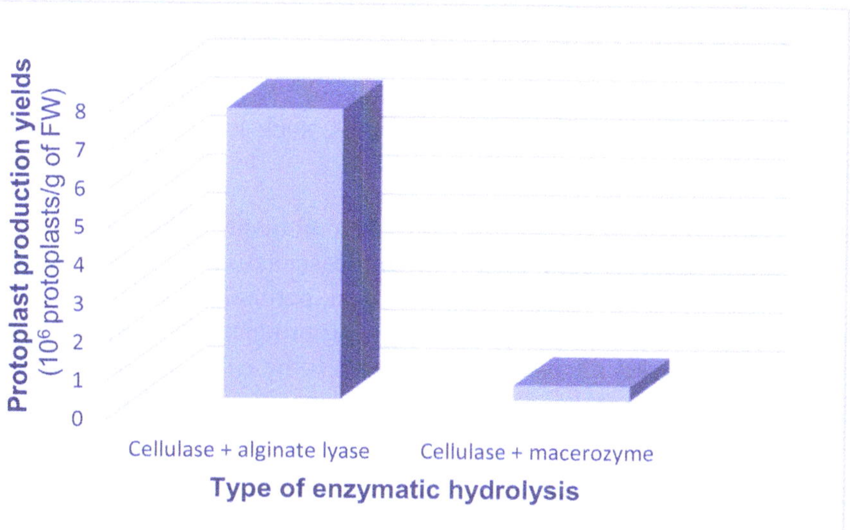

Figure 3.16. *Comparison of the efficacy of the combined use of cellulase + alginate lyase (cellulase Onozuka RS + alginate lyase) and cellulase + pectinases (cellulase Onozuka RS + Macerozyme R-10) in the production of protoplasts of* Undaria pinnatifida *(according to Avila-Peltroche et al. (2022))*

The development of a process for the enzymatic hydrolysis of algal cell walls, and above all its optimization, requires a sound knowledge of the composition and parietal structure. This last aspect, along with the availability of enzymes, remains a fundamental stumbling block in enzymatic engineering applied to the production of protoplasts.

Many of the enzymes used in these processes are of commercial origin or extracted from organisms. This is particularly true of alginate lyase, which is obtained from the digestive juices of grazing gastropods such as abalone. However, the collection and use of these digestive juices, and the purification of the enzymes involved, are not easy to implement given the limited accessibility and availability of this animal resource. Other biological sources can be used to obtain this type of enzyme. These include marine bacteria, but their isolation and cultivation require very specific microbiological practices.

To overcome this difficulty, the use of recombinant enzymes derived from genetic engineering is a possible solution. The feasibility of producing *Laminaria japonica* protoplasts using recombinant alginate lyase has been demonstrated (Inoue et al. 2008). However, this process requires the combined use of other enzymatic activities, such as cellulase and protease activities.

In this context, the use of recombinant alginate lyase and β-1,4-endoglucanase combined with protease activity enables protoplasts (5×10^6 protoplasts/g of MF) to be obtained, particularly from the epidermal zone of the thallus (Inoue et al. 2011). The production of protoplasts from this zone is of great interest, as it is known to be easily regenerable and the cells derived from it have a significant capacity to regenerate thalli and therefore sporophytes. The use of recombinant enzymes for the production of protoplasts from *L. japonica*, the world's main cultivated seaweed, can therefore be a biotechnological tool for the establishment of a reserve stock of genetically identical seaweeds.

The isolation of protoplasts is also applied to red algae (*Chondrus crispus*, *Porphyra* sp.). In species of the genus *Porphyra*, the main parietal polymer is a complex sulfated carbohydrate polymer called porphyran. This polysaccharide is structurally similar to agarose. In the alga *Porphyra linearis*, protoplast production is made possible by the use of a partially purified extract of agarases produced by the marine bacterium *Pseudomonas atlantica* (Chen et al. 1994). This extract contains two agarolytic enzymes known as β-agarase I and β-agarase II. The activity of the agarolytic extract increases with concentration until a plateau effect is observed at an enzyme concentration of 0.025% (w/v) (see Figure 3.17).

In contrast to the brown alga *L. japonica*, protoplast production from *P. linearis* is possible using only one type of hydrolytic activity, namely that of agarase. The use of abalone enzyme extracts, Macerozyme and cellulase extracts results in the isolation of non-viable protoplasts after 24 hours. These are referred to as green protoplasts as they have lost the pigment characteristic of red algae, phycoerythrin.

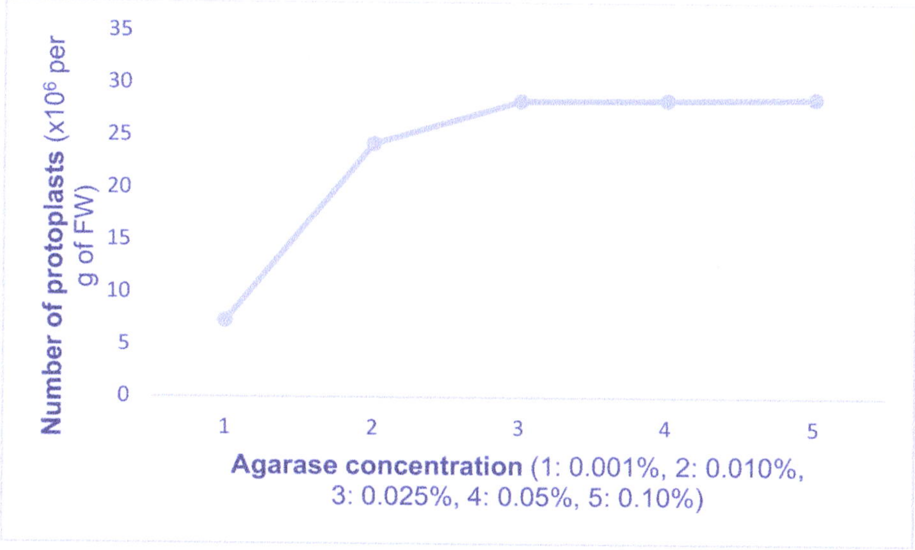

Figure 3.17. *Effect of different concentrations of agarases extracted from* Pseudomonas atlantica *on protoplast production from the alga* Porphyra linearis *after 4 h incubation (from Chen et al. (1994))*

The isolation of protoplasts by enzymatic engineering is not limited to the production of biological starting material useful for algal cultivation. This approach is also applicable to the use of protoplasts as a resource or model for studying the algal protein fraction. Indeed, the algal cell wall is a physical barrier that is highly resistant to conventional mechanical grinding methods, and the presence of large quantities of neutral or anionic parietal polysaccharides severely limits protein extraction (Fleurence 1999a).

In this context, protoplasts, cells stripped of their walls, are a biological material facilitating access to and study of the algal protein fraction. This type of approach has been tested on various algal species belonging to the main botanical groups (green algae, brown algae and red algae)

(Amano and Noda 1992). Protoplasts are produced from thallus fragments previously treated with protease A before being subjected to enzymatic cell wall digestion.

For green algae, this preparation includes cellulose Onozuka R-10 and Macerozyme R-10. This enzymatic blend is complemented by the use of an intestinal juice from the abalone *Haliotis discus* for the digestion of brown algae. For red algae, *Kallymenia perforata*, a mixture of abalone digestive juice and Macerozyme R-10, is used to isolate protoplasts.

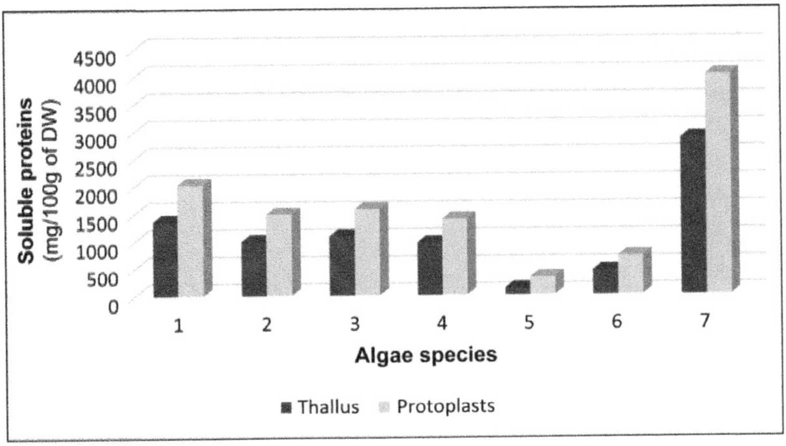

1: *Ulva pertusa*; 2: *Ulva fasciata*; 3: *Enteromorpha linza*; 4: *Monostroma nitidum*; 5: *Laminaria japonica* 6: *Undaria pinnatifida* 7: *Callymenia perforata*

Figure 3.18. *Comparison of the extraction of the soluble protein fraction between thallus and protoplasts of different algal species (from Amano and Noda (1992)). For a color version of this figure, see www.iste.co.uk/fleurence/aquatic.zip*

The production of protoplasts significantly improves the extraction yield of the soluble protein fraction compared with the yield obtained by mechanical grinding of the thallus (see Figure 3.18). An improvement in extraction yield of at least 40% is generally observed with this process. This approach can also be applied to the red alga *Porphyra yezoensis*, which is used in human nutrition under the name of nori and the total protein content of which can reach up to 47% of the seaweed's dry matter (Fleurence 1999b). In the case of this species, the increase in the extraction yield of the soluble protein fraction by enzymatic wall hydrolysis is close to 30% (Amano and Noda 1990).

Algal wall hydrolysis therefore appears to be an interesting biotechnological tool for improving the accessibility of water-soluble proteins. In this context, the use of polysaccharidases to improve protein solubilization from algae with high protein concentrations (>20% dry weight) has been the subject of numerous studies. In the red algae *Palmaria palmata*, *Chondrus crispus* and *Gracilaria verrucosa*, enzymatic digestion of the thallus by enzymes specific to parietal compounds has produced results similar to those previously described for protoplast production. The enzymes used are purified and specific to the main polysaccharide contained in the wall of each species (see Table 3.5) (Fleurence et al. 1995). The walls of the algae concerned also contain cellulose, which is considered a minor component by weight (3%) but has a major structural role according to some algal parietal models (Kloareg and Quatrano 1988).

This context therefore involves testing the effect of cellulase alone and in combination with other enzymes on wall degradation and on improving the extractability of soluble proteins. In the procedure used, the control experiment consists of incubating the algae in the incubation solution without enzymes and therefore without grinding the thallus. This latter aspect distinguishes the procedure used from those previously described.

Algal species	**Main enzyme tested**	**Specific activity (U.mL)$^{-1}$**	**Enzyme combination tested**
Palmaria palmata	Xylanase	46.5	Xylanase + cellulase
Chondrus crispus	Carrageenase	390	Carrageenase + cellulase
Gracilaria verrucosa	Agarase	2,432	Agarase + cellulase

Table 3.5. *Characteristics of the enzymes tested on the thalli of different algae to improve the extraction of algal proteins (from Fleurence et al. (1995))*

Incubation of the algae thalli in an enzyme-free buffer solution allows protein extraction to begin. This extraction, without prior grinding or enzyme treatment, occurs mainly due to the diffusion of phycoerythrin, a protein highly present in algae and easily solubilized by osmotic shock (see Figure 3.19). The enzymatic treatment applied gives variable results depending on the species and combinations tested. In the agarophyte species

G. verrucosa, the use of agarase or cellulase alone did not significantly improve protein extraction. In contrast, the combined use of agarase and cellulase increased extraction of the water-soluble protein fraction by a factor of three. A synergistic effect of these enzymes on thallus degradation is thus observed. In the carrageenophyte *C. crispus*, a hydrolysis synergy with a positive effect on protein extraction was also noted when carrageenase and cellulase were used simultaneously.

In contrast, the combined use of xylanase and cellulase with the xylanophyte *P. palmata* showed an antagonistic effect on protein extraction (see Figure 3.19). A study carried out in 1992 revealed that the combined action of purified xylanase and cellulase did not improve the liquefaction process of the alga *P. palmata* indicative of parietal hydrolysis (Lahaye and Vigouroux 1992).

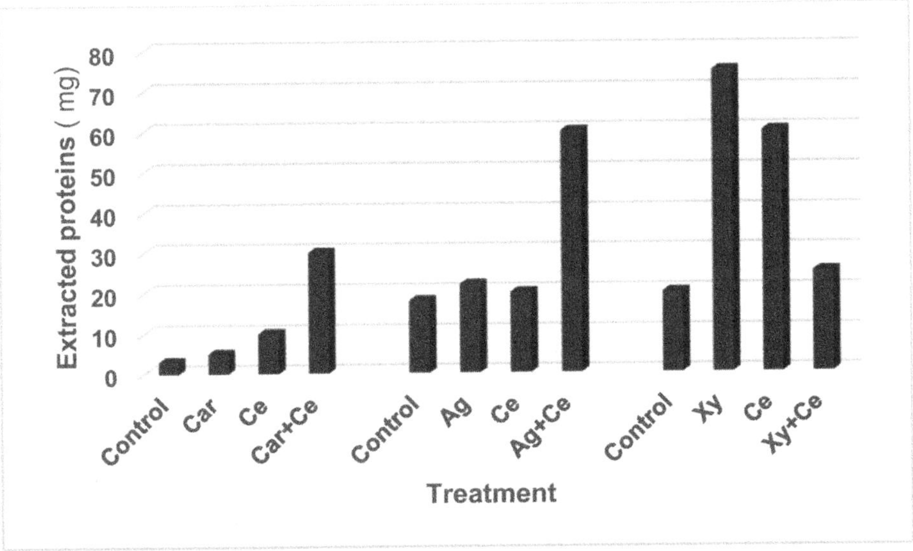

Chondrus crispus (Car: carrageenase; Ce: cellulase; Car + Ce: carrageenase + cellulase), *Gracilaria verrucosa* (Ag: agarase; Ce: cellulase; Ag + Ce: agarase + cellulase) and *Palmaria palmata* (Xy: xylanase; Ce: cellulase; Xy + Ce: xylanase + cellulase)

Figure 3.19. *Impact of enzymatic treatment on protein extraction from the algae* Chondrus crispus, Gracilaria verrucosa *and* Palmaria palmata *(from Fleurence et al. (1995))*

The results observed with *P. palmata* are obtained via the action of xylanases purified from fungal micro-organisms such as *Disporotrichum* or *Aspergillus niger*.

In contrast, the combined use of industrial enzyme preparations with xylanase activity and industrial preparations with cellulolytic activity does not induce antagonistic effects when treating *P. palmata* algae. This is the case for the combined use of Shearzyme, a xylanase produced by Novo via a genetically modified *Aspergillus oryzae* and the Celluclast preparation, a cellulase obtained from *Trichoderma reesei* (Joubert and Fleurence 2008). An activity unit (U) is defined as enzymatic activity releasing 1 µmol of monosaccharide per minute. The combined use of Shearzyme and Celluclast significantly improves the extractability of *P. palmata* proteins (see Figure 3.20). This improvement is greater the greater the concentration of Shearzyme used, the concentration of Celluclast used being constant in the combinations tested (12 units of cellulasic activity) (see Figure 3.20).

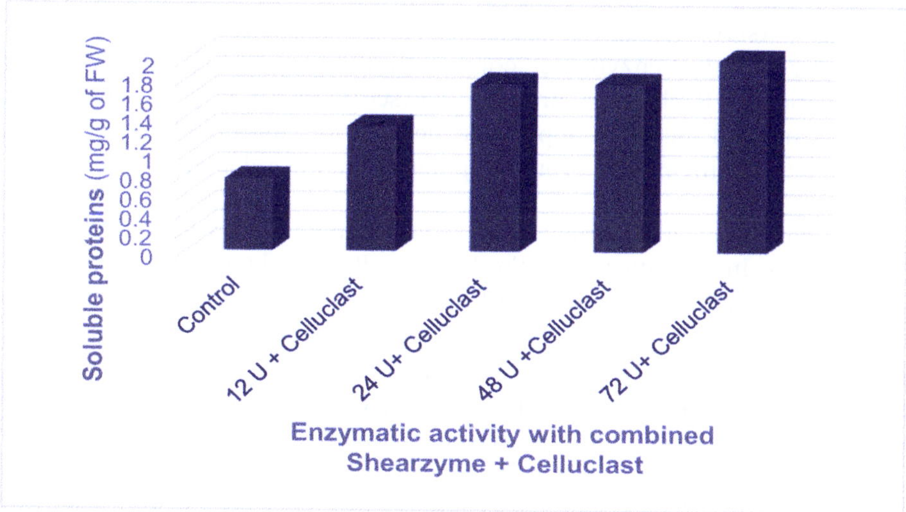

U: xylanase activity units

Figure 3.20. *Effect of the combination of Sheazyme (xylanase) and Celluclast (cellulase) on protein extraction from the red alga* Palmaria palmata *(from Joubert and Fleurence (2008))*

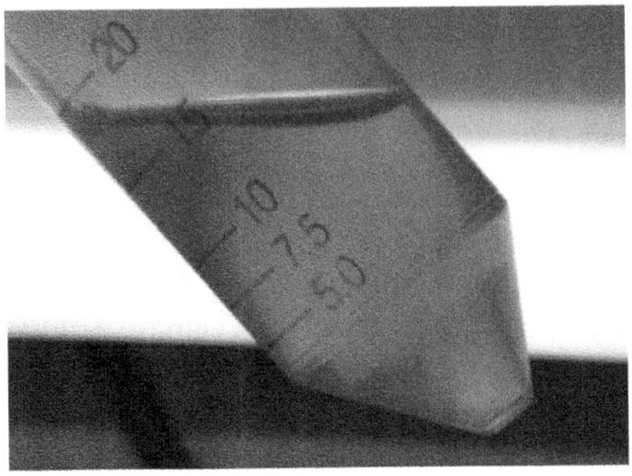

Figure 3.21. *R-Phycoerythrin from* Palmaria palmata *(photo credit © Fleurence J. 2006)*

The use of Shearzyme with Celluclast at a constant concentration (12 U) has also proved to be an effective method for extracting R-phycoerythrin. This purple protein pigment has a wide range of applications in the biomedical field (immunofluorescent reagent) and in the food industry (food coloring in Asia) (see Figure 3.21). The use of increasing concentrations of Shearzyme in the enzymatic combination significantly improves the solubilization of R-phycoerythrin (R-PE) (see Figure 3.22). The enzymatic digestion method therefore appears to be a gentle and effective extraction method for solubilizing this protein pigment from the red alga *P. palmata*.

This approach is also applicable to the extraction of *P. palmata* DNA (see Figure 3.23). In the latter case, the best results are obtained with xylanase concentration activity reduced to 12 units. In contrast to protein extraction, DNA solubilization is not improved when xylanase concentration is increased. It even decreases when xylanase activity is increased to 72 units (see Figure 3.22). This may be due to the release of more DNA-interacting oligosaccharides when hydrolytic activity is increased.

An analogous DNA extraction process is also possible for all algae, and particularly for red algae. This patented process involves a DNA purification method based on an algal enzymatic digestion step (see Figure 3.24) (Fleurence and Joubert 2006).

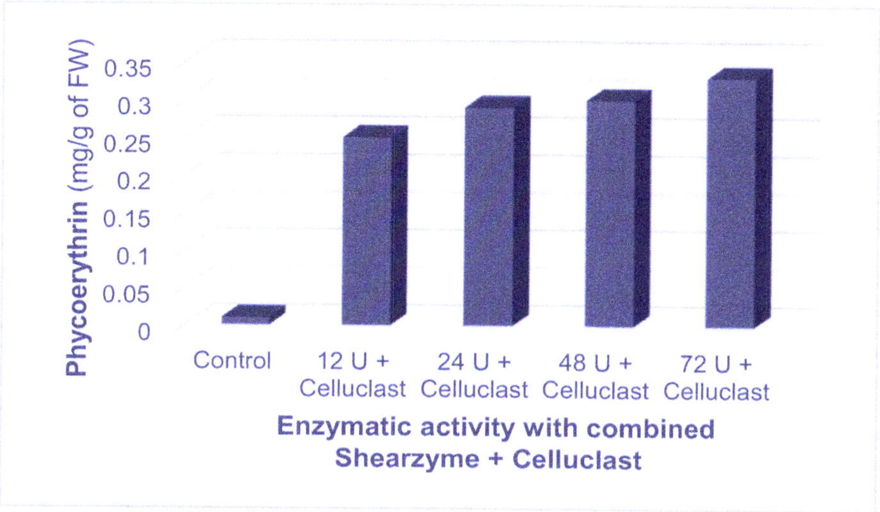

U: xylanase activity units

Figure 3.22. *Effect of Shearzyme enzymatic treatment (xylanase) and Celluclast (cellulase) on the solubilization of R-phycoerythrin from the red alga* Palmaria palmata *(from Joubert and Fleurence (2008))*

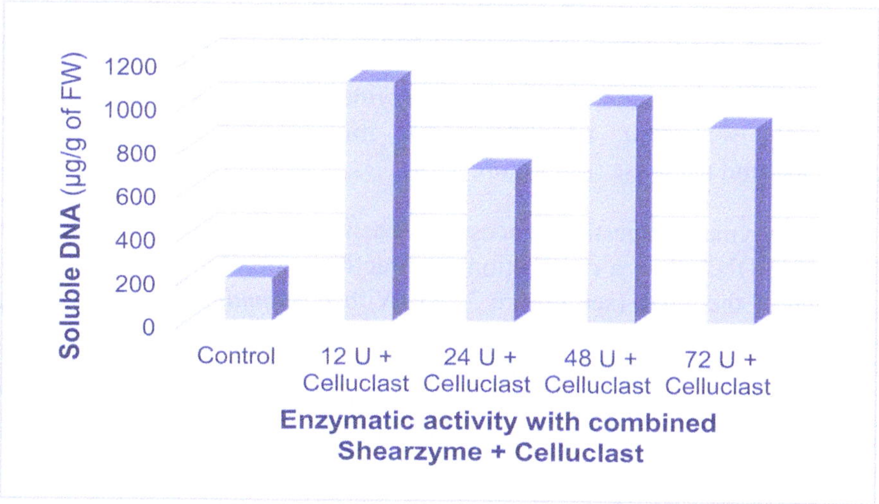

U: xylanase activity units

Figure 3.23. *Effect of Shearzyme enzymatic treatment (xylanase) and Celluclast (cellulase) on DNA solubilization from the red alga* Palmaria palmata *(from Joubert and Fleurence (2008))*

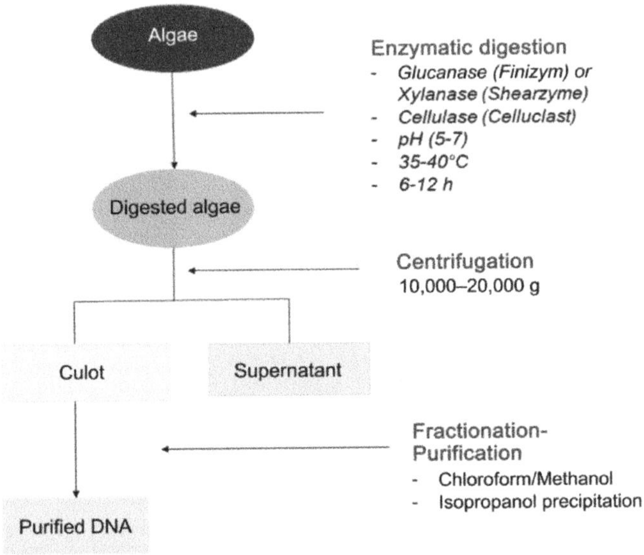

Figure 3.24. *Algal DNA purification process based on an enzymatic digestion extraction step (from Fleurence and Joubert (2006))*

This general process is adapted according to the species treated. For example, for the xylanophyte *P. palmata*, the enzyme preparation used is a mixture of xylanase and cellulase (Shearzyme + Celluclast). For the red alga *Gracilaria verrucosa*, agarophyte, the mixture used, is composed of glucanase and cellulase (see Figure 3.24).

The enzymatic digestion process has been proved to be more effective in extracting DNA than a conventional extraction process based on cryogenic grinding of the algae (see Figure 3.25). With *P. palmata* algae, this process improves the DNA recovery rate by almost 40% compared with a conventional process. The DNA recovered at the end of the purification process has the same degree of purity as that obtained via extraction by cryogenic grinding (purity index (PI) of 1.7). The enzymatic process therefore makes it possible to obtain DNA of comparable purity to that of the conventional process but with a very significant improvement in extraction yield. For *G. verrucosa* algae, the gain in the DNA extraction yield is less than that obtained with the conventional process (+20%). However, the PI of the DNA after application of this method is significantly higher than that observed with the conventional method (see Figure 3.26). It is also higher than that reported for *P. palmata*, whatever the extraction method applied to this alga.

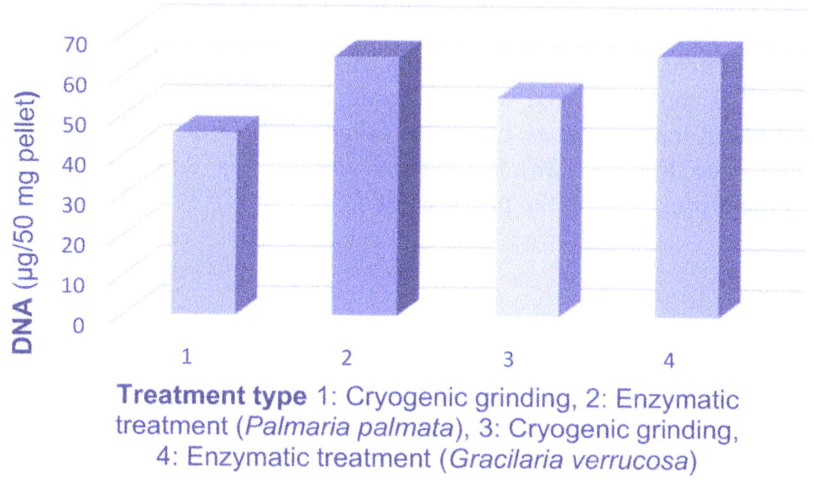

Figure 3.25. *Comparison of efficiency of mechanical grinding (cryogenic grinding) and enzymatic digestion in DNA extraction from the red algae* Palmaria palmata *and* Gracilaria verrucosa *(from Fleurence and Joubert (2006))*

Initially developed as a gentle protein extraction process, enzymatic digestion is therefore a suitable protocol for extracting algal DNA. In particular, this procedure enables the degradation of polysaccharides, which are potential contaminants of DNA and therefore inhibitors of its amplification by TAQ polymerase. This last point remains the main claim established by patent WO2006/016023.

The effectiveness of enzymatic maceration may also depend on the nature of the treatment undergone by the algal material prior to application of the enzyme preparations. For *P. palmata* algae, enzymatic hydrolysis shows very different efficiencies in terms of protein or R-phycoerythrin (R-PE) extraction depending on the procedure pre-applied to the thallus (cutting, grinding, cryogenic grinding) and the state of the starting material (fresh algae, dehydrated algae) (Dumay et al. 2013). Regardless of the pre-treatment process applied prior to enzymatic digestion, the best extraction results are obtained with fresh seaweed and not with freeze-dried seaweed. This is an interesting finding, as dehydration of algal material is generally a costly step in an industrial process. For fresh algae, the nature of the pre-treatment has a very significant influence on the extraction yield of R-PE after enzymatic digestion of the algae with xylanase (see Figure 3.27). Grinding, a classic and inexpensive process compared to cryogenic grinding, appears to be the best pre-treatment for extracting R-PE by enzymatic digestion of the thallus. The picture is more nuanced when it comes to extracting soluble proteins. Indeed, thallus cutting is as effective as grinding as a pre-treatment for extraction (see Figure 3.28). Cryogenic grinding is less effective as a pre-treatment process than the previous two.

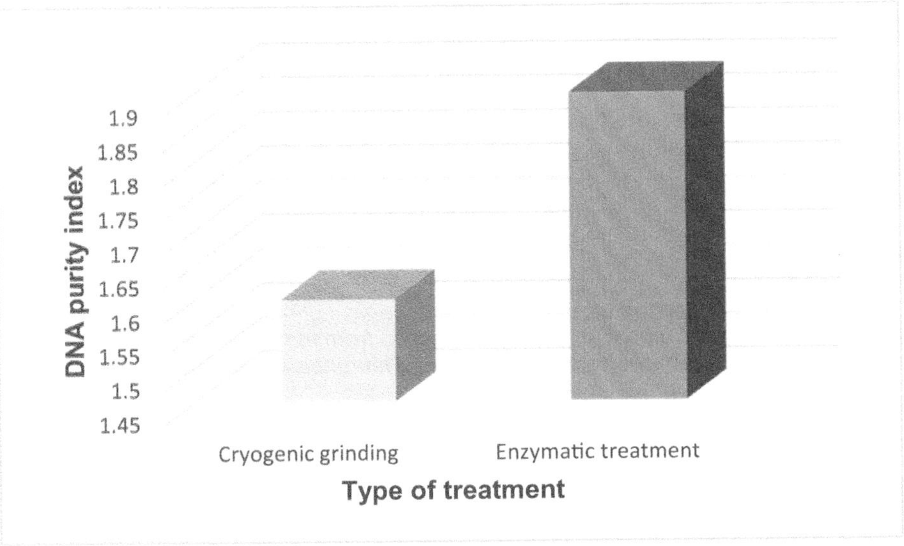

Figure 3.26. *Influence of extraction method (cryogenic grinding, enzymatic digestion) on the purity index of DNA obtained from the red alga* Gracilaria verrucosa *(from Fleurence and Joubert 2006))*

The nature of the resource, whether fresh or dehydrated, and the type of pre-treatment applied can have a significant influence on the efficiency of the enzymatic process used as a soft extraction method. This type of observation mainly concerns *P. palmata*, which remains a historical model for the study of enzymatic hydrolysis applied to the improvement of protein extraction, including R-PE.

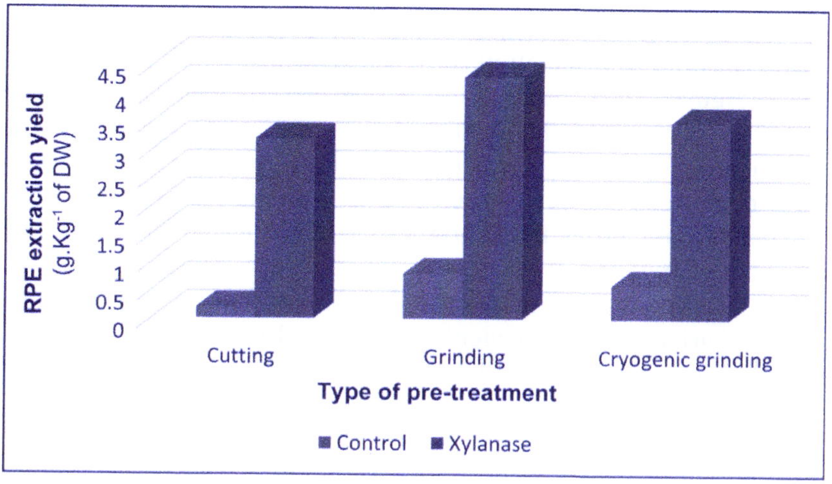

Figure 3.27. *Effect of pre-treatment of* Palmaria palmata *algal thallus (cutting, grinding, cryogenic grinding) on the extraction yield of R-phycoerythrin after enzymatic digestion with xylanase (from Dumay et al. (2013)). For a color version of this figure, see www.iste.co.uk/fleurence/aquatic.zip*

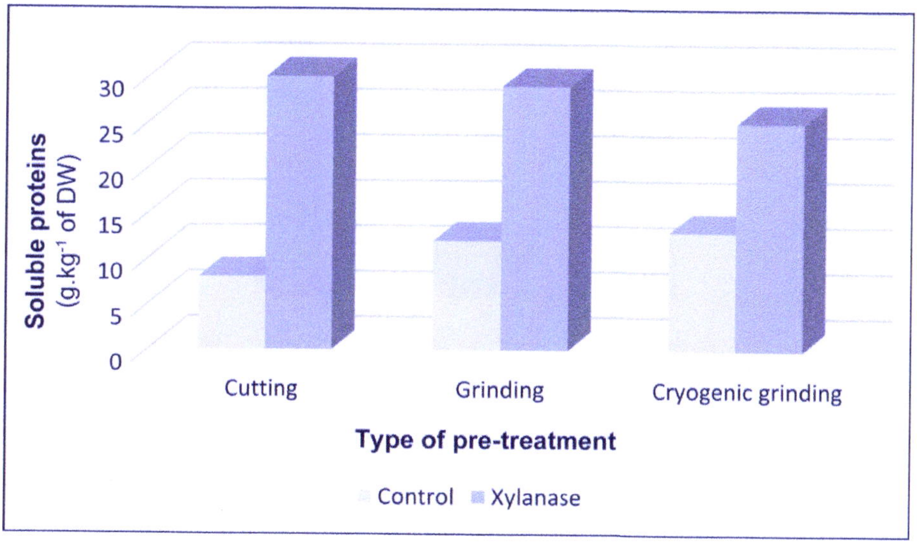

Figure 3.28. *Effect of pre-treatment of* Palmaria palmata *algal thallus (cutting, grinding, cryogenic grinding) on the extraction of soluble proteins after enzymatic digestion with xylanase (from Dumay et al. (2013)). For a color version of this figure, see www.iste.co.uk/fleurence/aquatic.zip*

With the red alga *Mastocarpus stellatus*, cryogenic grinding is used as a pre-treatment prior to maceration of freeze-dried thalli with β-glucanase and xylanase (Nguyen et al. 2017). Fresh seaweed, in contrast, undergoes a preliminary cutting of the thallus prior to enzymatic maceration. In terms of R-PE and soluble protein extraction, the best results are obtained with dehydrated seaweed subjected to cryogenic grinding (see Figures 3.29 and 3.30). These results differ from those reported for *P. palmata* algae, which shows that fresh raw material is better suited than dehydrated algal material to enzymatic maceration for extracting R-PE. The efficiency of pre-treatments (i.e. cutting, grinding and cryogenic grinding) also varies according to the species considered (*P. palmata, M. stellatus*).

The condition of the algal matrix and pre-treatments prior to maceration are therefore important elements to take into account when optimizing enzyme hydrolysis. These parameters may differ between species even if they are in the same botanical group (e.g. red algae).

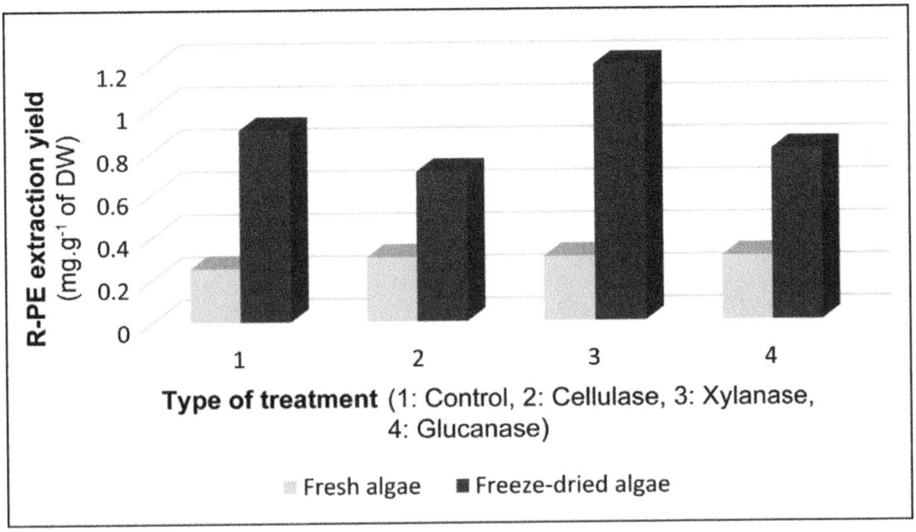

Figure 3.29. *Effect of enzymatic treatment on R-phycoerythrin extraction as a function of raw material pre-treatment (cutting for fresh algae and cryogenic grinding for freeze-dried algae) (from Nguyen et al. (2017)). For a color version of this figure, see www.iste.co.uk/fleurence/aquatic.zip*

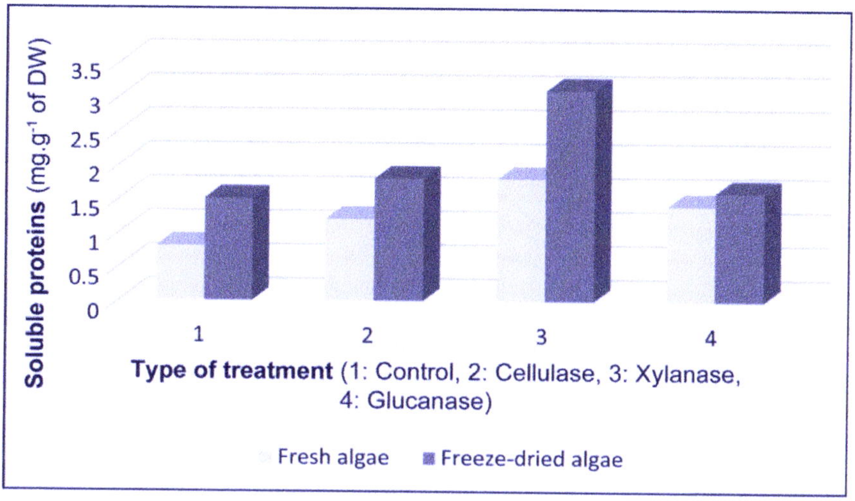

Figure 3.30. *Effect of enzymatic treatment on protein solubilization as a function of raw material pre-treatment (cutting for fresh algae and cryogenic grinding for freeze-dried algae) (from Nguyen et al. (2017)). For a color version of this figure, see www.iste.co.uk/fleurence/aquatic.zip*

Enzymatic hydrolysis as a useful biotechnological tool for the gentle extraction of algal proteins is therefore a process to be developed for each species and should not be generalized to a botanical group or family.

The enzymatic hydrolysis of algal thalli is not only applicable to the production of protoplasts or to improving the extraction efficiency of water-soluble molecules, such as proteins, pigments or DNA.

The enzymatic digestion of algal polysaccharides also presents an advantage in terms of adding value to algal resources. Indeed, algal polysaccharides often act as dietary fibers, inhibiting protein digestibility. Their main mode of action is the inhibition of digestive proteases such as pepsin, trypsin or chymotrypsin (Horie et al. 1995).

Brown algae such as *Laminaria japonica* (kombu), *Hizikia fusiforme* (hijiki) or *Undaria pinnatifida* (wakame) have soluble fibers that strongly inhibit pepsin (see Figure 3.31). Enzymatic maceration of these seaweeds using alginate lyase is an approach that can partially or totally remove this inhibitory effect on pepsin.

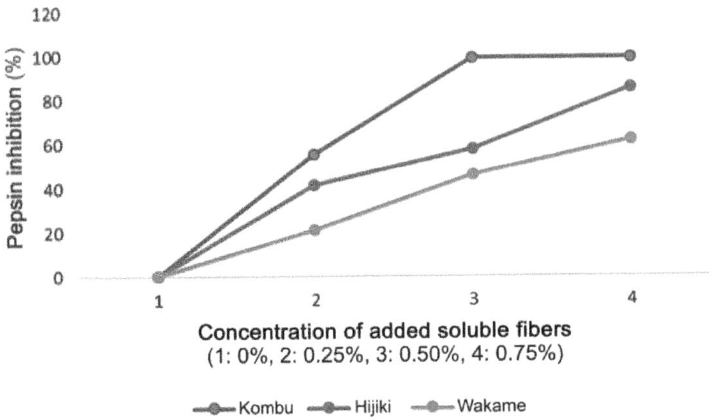

Figure 3.31. *Effect of adding soluble fiber extracted from* Laminaria japonica *(kombu),* Hizikia fusiforme *(hijiki) and* Undaria pinnatifida *(wakame) algae on the inhibition of the gastric protease pepsin (from Horie et al. (1995)). For a color version of this figure, see www.iste.co.uk/fleurence/aquatic.zip*

The impact of enzymatic maceration on the elimination of polysaccharides, an antinutritional factor limiting the digestibility of the algal protein fraction, is well-documented for the red alga *P. palmata*. This approach has been patented with PCT extension WO 02/07528 A1 (Fleurence et al. 2002). The enzymes used to degrade *P. palmata* polysaccharides are xylanases, cellulases and β-glucanases, which are supplied by manufacturers and derived from micro-organisms belonging to the genera *Trichoderma* or *Aspergillus* (see Table 3.6). The in vitro digestibility of *P. palmata* proteins is significantly improved by the combined use of xylanase and cellulase (see Figure 3.32). These enzymes specifically degrade xylans and cellulose. These two polysaccharides can therefore be considered antinutritional factors, and their enzymatic elimination improves the action of pepsin and pancreatin, the digestive enzymes used in in vitro studies of algal protein digestibility. Furthermore, the substitution of Celluclast with β-glucanases (Ultraflow, Finizym) in the enzyme combination tested gave poorer results than those obtained with the Shearzyme–Celluclast combination (see Figure 3.33).

Apart from the impact on protein digestibility, the use of enzymes, and more specifically enzyme combinations, can also have an effect on the quality of solubilized algal proteins. Indeed, the combined use of Shearzyme and Celluclast allows the recovery of a protein fraction with a methionine content, an essential sulfur-containing amino acid, representing 0.53% of the total amino acid fraction (see Figure 3.34). Similarly, the Shearzyme and Finizym combination was the most effective in recovering a threonine-enriched protein fraction (see Figure 3.34). This is an interesting finding, as it shows that the enzymatic hydrolysis process is effective in improving the digestibility of the algal protein fraction but can also be adapted to the preferential extraction of proteins with different essential amino acid profiles.

Enzymes	Origins	Suppliers
Cellulases Celluclast Cellulyve	*Trichoderma reesei*	Novo Lyven
Xylanases Shearzyme Panlyve	Genetically modified *Aspergillus oryzae* *Aspergillus niger*	Novo Lyven
β-Glucanases Finizym Ultraflow	*Aspergillus niger* *Humicola insolens*	Novo

Table 3.6. *Examples of enzymes used for hydrolysis of* Palmaria palmata *polysaccharides (xylans, cellulose) to improve in vitro digestion of the algal protein fraction (Fleurence et al. 2002)*

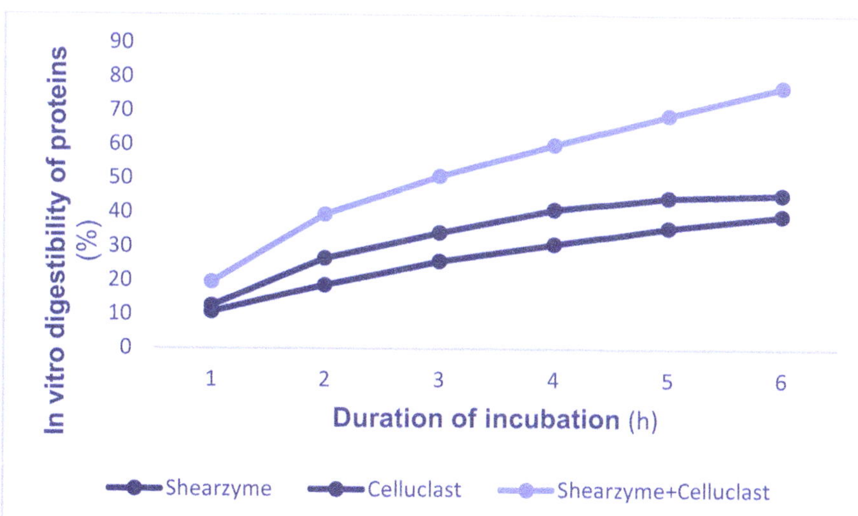

Figure 3.32. *Effect of isolated or combined action of xylanases (Shearzyme) and cellulases (Celluclast) on in vitro protein digestibility with* Palmaria palmata *(from Fleurence et al. (2002)). For a color version of this figure, see www.iste.co.uk/fleurence/aquatic.zip*

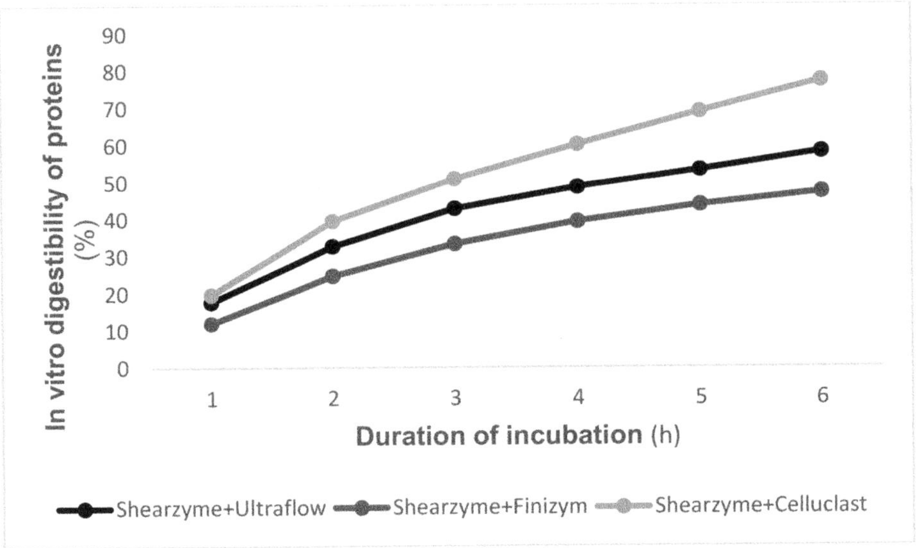

Figure 3.33. *Effect of the combined action of xylanases (Shearzyme) and cellulases (Celluclast) or β-glucanases (Ultraflow, Finizym) on the in vitro digestibility of* Palmaria palmata *proteins (from Fleurence et al. (2002)). For a color version of this figure, see www.iste.co.uk/fleurence/aquatic.zip*

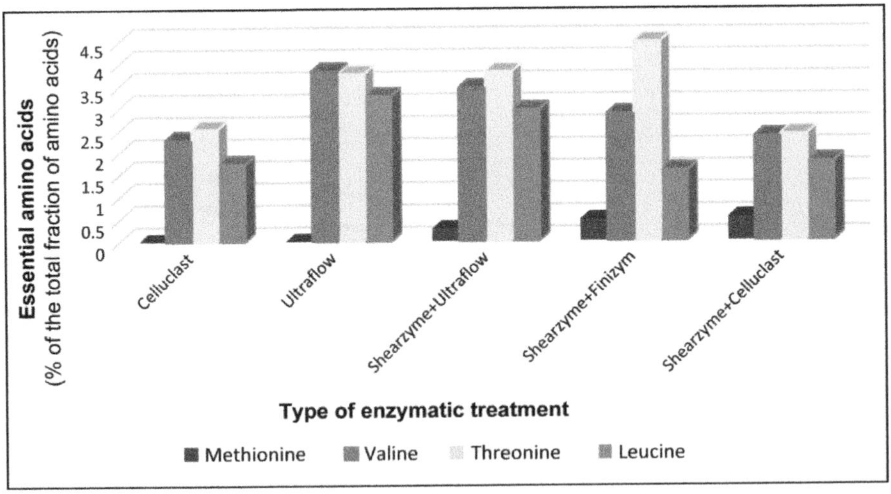

Figure 3.34. *Effect of enzymatic hydrolysis via xylanases (Shearzyme), cellulases (Celluclast) or β-glucanases (Ultraflow, Finizym) on the relative amino acid content of extracted proteins (from Fleurence et al. (2002)). For a color version of this figure, see www.iste.co.uk/fleurence/aquatic.zip*

Enzymatic hydrolysis of the macroalgal matrix is a process with multiple applications (protoplast production, extraction of high-value-added molecules, elimination of antinutritional factors). It can also be used to extract polysaccharides such as ulvans, the biological activities of which (anticancer, anti-inflammatory, antibacterial and antiviral) are of major interest in terms of therapeutic applications (Cindana Mo'o et al. 2020).

The use of polysaccharidase-type hydrolytic enzymes is also possible with a microalgal matrix. Applications mainly involve the extraction of molecules of industrial interest (proteins, pigments, oligosaccharides, etc.) or the production of fermentable sugars.

However, some microalgae are characterized by a highly complex composition and parietal structure, limiting the efficiency of the enzymatic hydrolysis process. Despite this, it is possible, with many species of green algae, to use cellulases and amylases to degrade cell wall (cellulose) and storage (starch) polysaccharides. This approach, known as saccharification, allows the production of simple sugars, potential substrates for subsequent fermentation (Fleurence 2021a).

Whatever the algal matrix, the enzymatic approach requires a good knowledge of the nature of the parietal polysaccharides and the appropriate enzymes. These two aspects remain the main limiting steps in enzymatic engineering applied to algae.

3.2. Fermentation

Fermentation is considered a traditional biotechnological process. Broadly speaking, it can be defined as the transformation of an organic matrix by enzymes produced by micro-organisms (fungi, yeasts, bacteria). Food fermentation is an age-old process used for food preservation.

3.2.1. *Fermentation applied to fish matrices*

Traditionally, fermentation is used to improve the preservation of fish, a food whose shelf life can be very limited. One of the most popular fermented products is fish sauce, where the process begins with the action of endogenous proteases (e.g. calpains) and continues through a fermentative stage (see section 3.1.1). In addition, other products are obtained from the fermentation of fish, and particularly herring. In Northern Europe, these traditional products are known as rakeorret (Norway), surströmming (Sweden) and maatjes (the Netherlands) (Marti-Quijal et al. 2020a). Surströmming is a fermented product made from washed and salted herring. The fermenting herring is then stored in closed cans where the fermentation process continues. The first stage of fermentation is achieved by autolysis of the flesh via endogenous proteases. The second stage relies on the action of fermentative bacteria of the genus *Haloanaerobium* (Kobayashi et al. 2000). The fermentation process produces various organic acids, which give the final product its characteristic odor (see Figure 3.35). Butyric acid, produced by the fermentation process, is the source of the rancid butter odor characteristic of surströmming. The pungent smell, in contrast, is generated by the production of propanoic acid via the fermentation process.

In addition to herring, other fish species can also be transformed via a fermentative process (see Table 3.7) (Skara et al. 2015). These include sharks in Iceland (hakari) and trout or char in Norway (rakfisk).

Bacteria of the genus *Lactobacillus* are present in all these traditional fermentary processes (see Table 3.7).

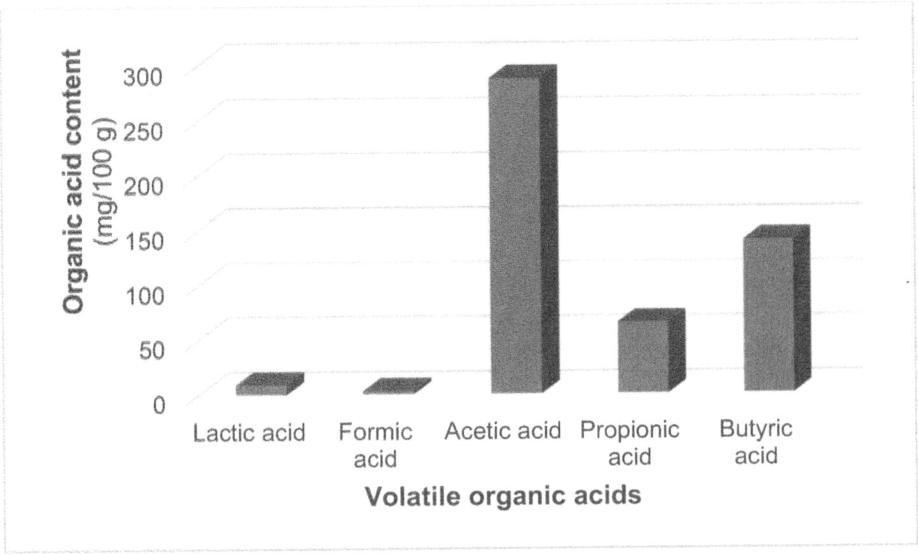

Figure 3.35. *Volatile organic acids produced by fermentation during the production of surströmming (from Kobayashi et al. (2000))*

In Southern Europe, *anchoitage* is a ripening process applied to anchovies. It involves two distinct stages. The first involves pre-salting the fresh anchovies with brine. Second, the fish are placed in barrels with salt in alternating layers (fish/salt). Once the maturing process is complete, the salt is rinsed off and the fish covered with oil and packed in jars.

Species	Fermentation process	Main micro-organisms involved	Name of fermented product	Origin
Greenland shark (*Somniosus microcephalus*)	Fish are buried in pits lined with stones or algae for several weeks or months Fermentation in soil	*Moraxella* spp. *Acinetobacter* spp. *Lactobacillus* spp.	Hakari	Iceland
Salmonids (*Salmo trutta*, *Salvelinus alpinus*)	Storage of eviscerated fish in watertight containers filled with salt water and kept at low temperature (3–7°C) for 3–12 months Fermentative maturation during storage	*Lactobacillus* spp.	Rakfisk	Norway

Table 3.7. *Examples of traditional fermentation processes and products in Northern Europe (from Skara et al. (2015))*

Anchoitage is an endogenous fermentation process based on internal proteolysis involving the proteases and on external proteolysis linked to the action of bacterial flora. The importance of this endogenous bacterial flora varies according to the origin of the anchovies (Atlantic Ocean, the Mediterranean). In anchovies from the Mediterranean, the level of the bacterial flora is subject to little variation. This is not the case with Atlantic anchovies, for which the microflora increases during the first nine weeks of maturation and then decreases. Whatever the origin of the anchovies, a decrease in non-halophytic flora in favor of halophytic and then hyperhalophytic flora is observed during maturation (Campello 1983). Hyperhalophytic flora is characterized principally by the presence of archaea. These archaea multiply for five to six weeks and then lyse. Complete maturation takes place after 16–17 weeks.

Ultimately, the organoleptic quality of the final product depends on the main criteria: the geographical origin of the fish, the importance of the associated bacterial flora and its affinity with salt. The last two criteria depend on the length of maturation.

The *anchoitage* process can be applied to a number of species worldwide. These include *Engraulis encrasicolus* (European anchovy), *Engraulis ringens* (Peruvian anchoveta), *Engraulis japonicus* (Japanese anchovy) and *Engraulis anchoita* (Argentine anchoita).

In Africa, fermentation applied to the preservation of fish products is a traditional process. The products processed in this way are known as adjuevan (Ivory Coast), lanhouin (Benin) and guedj (Senegal) (Fall et al. 2019). During the process, the fish undergo a double fermentation process:

– initial fermentation or primary fermentation in the absence or presence of water (see Figure 3.36). This stage is then followed by a salting phase, either dry or in brine;

– a spontaneous secondary fermentation process that takes place in a closed or open environment.

The fermented fish are then sun-dried to produce the final product, which is sold in local markets.

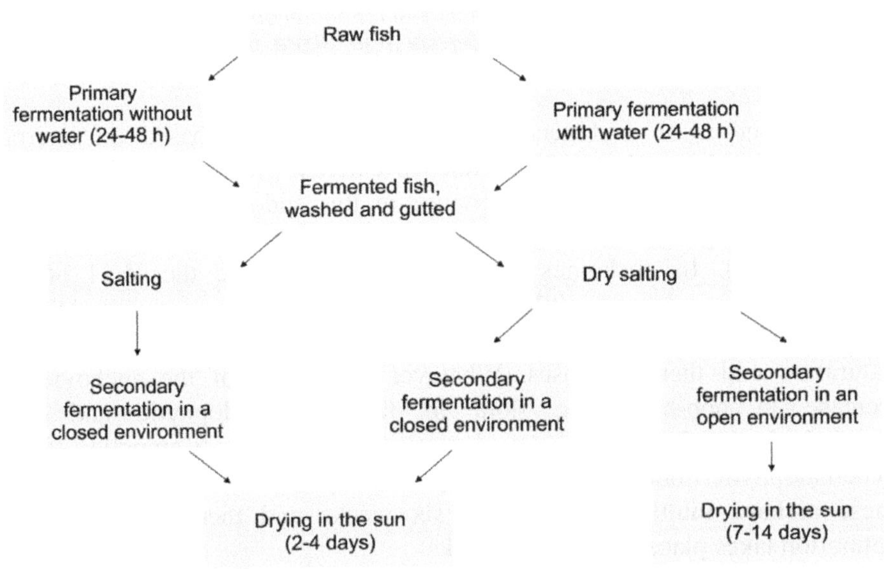

Figure 3.36. *Simplified fish fermentation process used in Africa to obtain the traditional adjuevan product (from Fall et al. (2019))*

This traditional process also raises research challenges. With regard to adjuevan, characterizing the bacteria and yeasts that predominate in fermentation is an important research objective. In particular, this could enable them to be used as ferments and make it possible to optimize or even standardize a traditional process that is still largely artisanal.

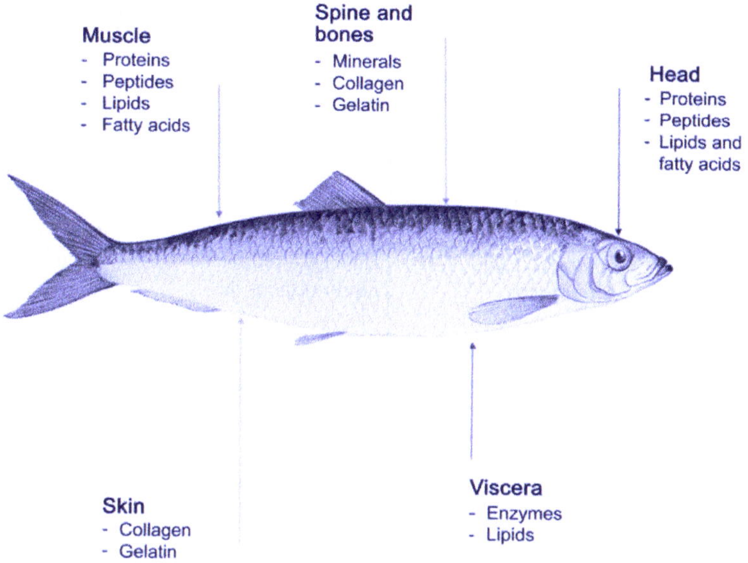

Figure 3.37. *Examples of valorization of fish co-products (from Marti-Quijal et al. (2020a))*

Fermentation applied to fish matrices is not limited to the implementation of food preservation processes. Indeed, the valorization of fish co-products (head, viscera, backbone and skin) (see Figure 3.37) is a major challenge for the fish and aquaculture processing industry.

In this context, the use of fermentative micro-organisms appears to be a useful biotechnological approach to the valorization of co-products. Fermentation realized by lactic acid bacteria is a safe process from a sanitary point of view and enables the recovery of bioactive compounds from fish skin, fish head viscera or fish silage (see Table 3.8). These compounds can display antioxidant or antimicrobial activities (Khubber et al. 2022). This is particularly true of the fermentation of sea bass (*Dicentrarchus labrax*) by the *Lactobacillus reuteri* strain, which leads to the production of reuterin

(Marti-Quijal et al. 2020a). This molecule is an antibiotic effective against Gram-positive and Gram-negative bacteria, as well as the yeast *Saccharomyces cerevisiae* and the protozoan *Trypanosoma cruzi*.

Resources	Fermentative bacteria or fungi	Bioactive compounds	Potential added value
Silage of fish	*Lactobacillus plantarum* *Lactobacillus brevis* *Pediococcus acidilactici* *Enterococcus gallinarum*	Peptides	Antimicrobial Antioxidant
Fish sauce	*Lactobacillus acidophilus*	HMG-Co reductase inhibitor peptide	Inhibitor of precursors of the cholesterol synthesis pathway
Muscle	*Lactobacillus curvatus* *Enterococcus faecium*	Preservation compounds	Conservation
Skin, viscera, head, fins	*Aspergillus oryzae* *Aspergillus niger* *Lactobacillus plantarum*	Odorous compounds Phenolic acids	Aroma
Loins	*Lactobacillus reuteri*	Reuterin	Antibiotic
Scales, intestine	*Bacillus subtilus*	Bioplastic	Packaging
Fish waste	*Lactobacillus pentosus*	Lactic acid	Packaging

Table 3.8. *Examples of applications of the fermentative process to fish co-products and associated potential uses (from Marti-Quijal et al. (2020a) and Khubber et al. (2022))*

Certain organs considered as co-products can also be used as a source of fermenting micro-organisms. This is particularly true of the digestive system, which is naturally colonized by lactic acid bacteria. In sea bass (*Dicentrarchus labrax*), some 30 lactic strains can be isolated from the stomach, small intestine and colon. Among these, seven belonging to the genus *Lactobacillus* have high proteolytic activity. These strains can be used to ferment meat extracts and co-products (head, bones) (see Figure 3.38) (Marti-Quijal et al. 2020b).

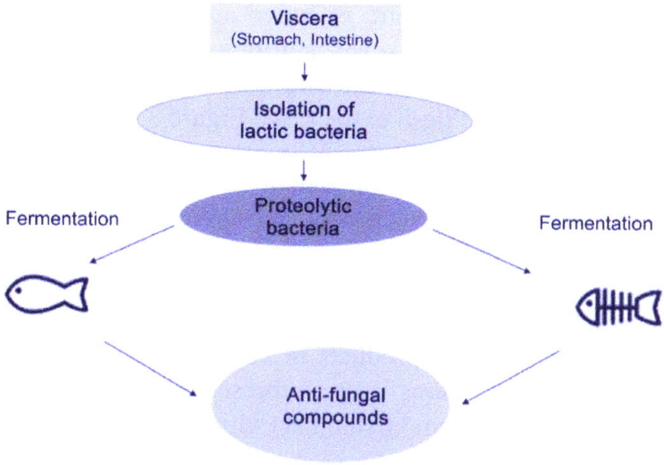

Figure 3.38. *Protocol for using lactic acid bacteria from sea bass viscera (*Dicentrarchus labrax*) in a fermentation process with flesh and co-products to obtain antifungal substances (from Marti-Quijal et al. (2020b)).*

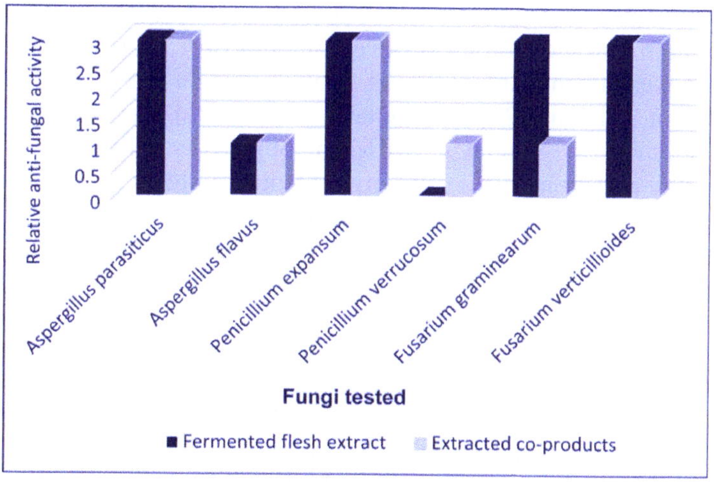

Figure 3.39. *Antifungal activities isolated from meat extracts or co-products fermented via the action of lactic acid bacteria from sea bass (*Dicentrarchus labrax*) viscera (from Marti-Quijal et al. (2020b))*

Some strains, particularly those originating in the stomach, generate a type of fermentation that produces compounds with antifungal activity (see

Figure 3.39). These substances are effective against the genera *Fusarium*, *Aspergillus* and *Penicillium*. However, antifungal efficacy differs according to the fungal species targeted and the origin of the fermented material. It is often higher with fermented flesh extract (see Figure 3.39).

Adding value to fishery and aquaculture by-products via fermentation appears to be a promising biotechnological approach. From an economic, ecological and societal point of view, it is becoming a necessity for many producer countries. By way of example, Norway, the world leader in salmonid aquaculture and a major player in the fishing industry, produced 401,000 tons of fish-based co-products in 2016 (Marti-Quijal et al. 2020a). Of this volume, 145,000 tons were represented by viscera, the main co-product from aquaculture and fishing activities.

3.2.2. *Fermentation applied to algal matrices*

Fermentation is a transformation process that can also be applied to algae. It is mainly used with macroalgae, which represent the main algal resource exploited for industrial applications. This approach has been supported by a major research and technological development effort over the last 20 years. Indeed, we went from 5 international publications on the subject in 2000 to almost 100 in 2019 and 2020 (Reboleira et al. 2021). Fermentation applied to algae pursues two somewhat different objectives. The first consists of improving the nutritional value and organoleptic qualities (texture, aroma) of the "vegetable" seaweed or intermediate food product (IFP). The second is to produce high-value-added molecules such as ethanol, lactic acid or bioactive peptides (see Figure 3.40).

The efficiency of the process often relies on a preliminary step called saccharification. During this stage, the algae are subjected to the action of cellulases or β-glucanases which break down the walls and their polysaccharides into simple sugars. These sugars or fermentable substrates are then transformed into lactic acid and ethanol by lactic acid bacteria and yeast (see Figures 3.40 and 3.41). Fermentation is a process which, depending on conditions, can start spontaneously (see Table 3.9). As a general rule, it is induced by the addition of a microbial inoculum (bacteria, yeast) to previously saccharified algal material. This inoculum is mainly made up of lactic acid bacteria belonging to the genus *Lactobacillus*. During the fermentation process, the species *L. brevis*, *L. casei* and *L. plantarum* rapidly become dominant in the lactic flora (Uchida and Miyoshi 2013).

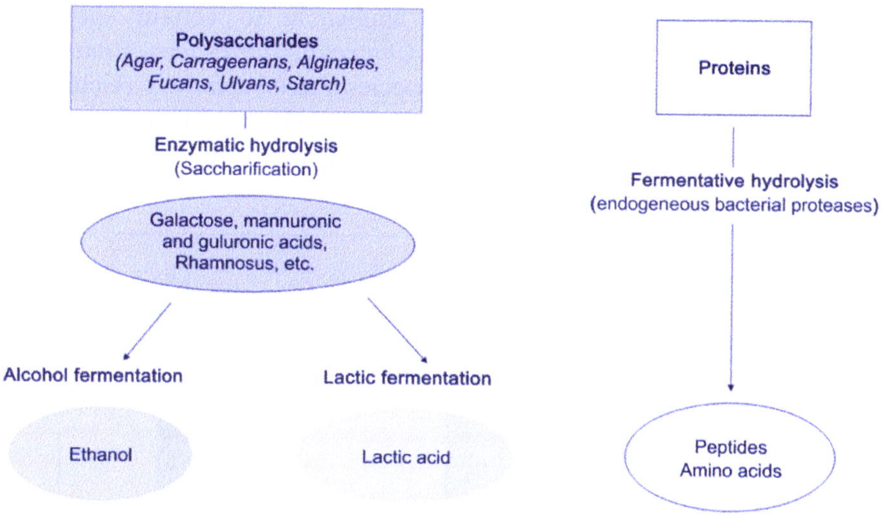

Figure 3.40. *Fermentation process for upgrading algal polysaccharides and proteins (from Reboleira et al. (2021))*

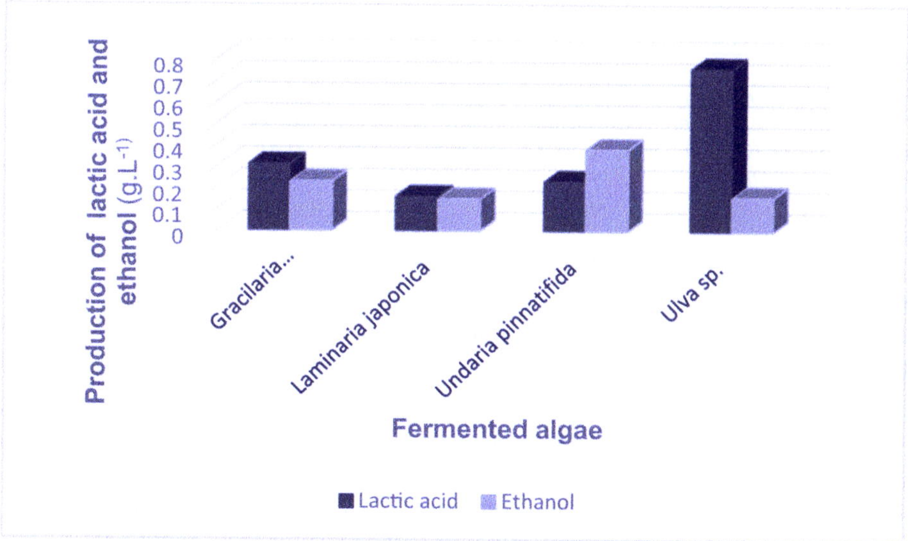

Figure 3.41. *Production of lactic acid and ethanol by fermentation of macroalgae via lactic acid bacteria (from Uchida and Miyoshi (2013)). For a color version of this figure, see www.iste.co.uk/fleurence/aquatic.zip*

Lactic acid fermentation is also applicable to certain species of microalgae. These belong to the genera *Chlorella*, *Tetraselmis*, *Chaetoceros* and *Nannochloropsis*, as well as the species *Pavlova lutheri* (Uchida and Miyoshi 2013). Fermentative lactic acid production is significantly enhanced by adding a macerozyme/lactase enzyme cocktail to cellulase during saccharification treatment (see Figure 3.42).

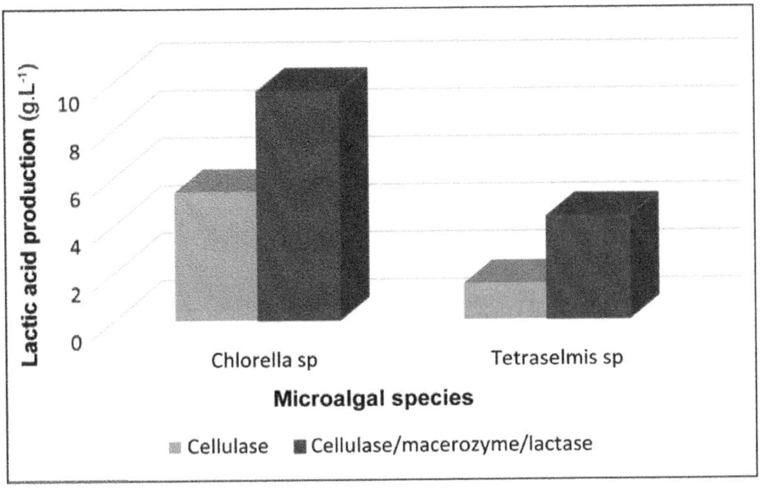

Figure 3.42. *Production of lactic acid by fermentation of microalgae (*Chlorella *sp.,* Tetraselmis *sp.) after pre-treatment with saccharifying enzymes (from Uchida and Miyoshi (2013)). For a color version of this figure, see www.iste.co.uk/fleurence/aquatic.zip*

During fermentation, algal polysaccharides are preferred substrates for the fermentative process. Some can be transformed by spontaneous fermentation, while others are degraded by the action of exogenous yeasts, fungi or lactic bacteria (see Table 3.9). Genetically modified micro-organisms are also sometimes used as part of this biotechnological approach. The products obtained from fermentation of sugars are ethanol or lactic acid, depending on the type of fermentation (see Figure 3.40). Hydrogen, an essential compound for the development of biofuels, is also a product generated by the fermentation of algal polysaccharides (see Table 3.9).

Proteins can also be hydrolyzed into peptides via a fermentative process. This process relies on the activation of the endogenous bacterial flora and its proteases (see Figure 3.40). It can also be enhanced by the use of exogenous proteases.

Polysaccharides	Micro-organisms	Valorized products
Agar	Spontaneous fermentation	Hydrogen
Carrageenan	*Saccharomyces cerevisiae*	Ethanol
Xylanes	Genetically modified *S. cerevisiae* Lactic acid bacteria	Oligosaccharides Lactic or acetic acid
Alginates	Spontaneous fermentation Genetically modified *Sphingomonas* sp. *Clostridium beijerinckii*	Hydrogen Ethanol Butanol, acetone, ethanol, butyrate
Ulvanes	Genetically modified *S. cerevisiae*	Ethanol

Table 3.9. *Some examples of fermentations applied to macroalgal polysaccharides for the production of high-value-added molecules (according to Reboleira et al. (2021))*

Apart from the production of high-value-added metabolites, fermentation remains a process associated with agri-food valorization. The use of macroalgae as fermentative substrates is currently an industrial issue involving the food, feed and agricultural sectors. As with other food products, fermentation of algae can lead to the production of lacto-fermented foods, beverages and fermented juices (see Figure 3.43) (Monteiro et al. 2021). Fermented juice from *Porphyra* sp. (*Susabinori*) is characterized by an estimated amino acid content of 165 g per $N.L^{-1}$ of liquid. By way of comparison, fermented soy sauce or Koikuchi sauce has an amino acid content of around 120 $g.L^{-1}$. Obtaining fermented seaweed juices as a source of amino acids for human consumption therefore represents a new opportunity to exploit the fermentative process applied to macroalgae.

In animal feed, fermented algae rich in oligosaccharide prebiotics and probiotic micro-organisms also appear to be a potentially promising application. Trials involving the use of algae silage in the feed of aquaculture-produced animals have produced encouraging results. Indeed, the introduction of 10% fermented *Ecklonia* into the diet of red sea bream significantly improves the resistance of the animal to iridovirus (see Figure 3.44). The virus responsible for iridovirus affects many species of fish farmed in aquaculture, including some highly commercial species such as sturgeon.

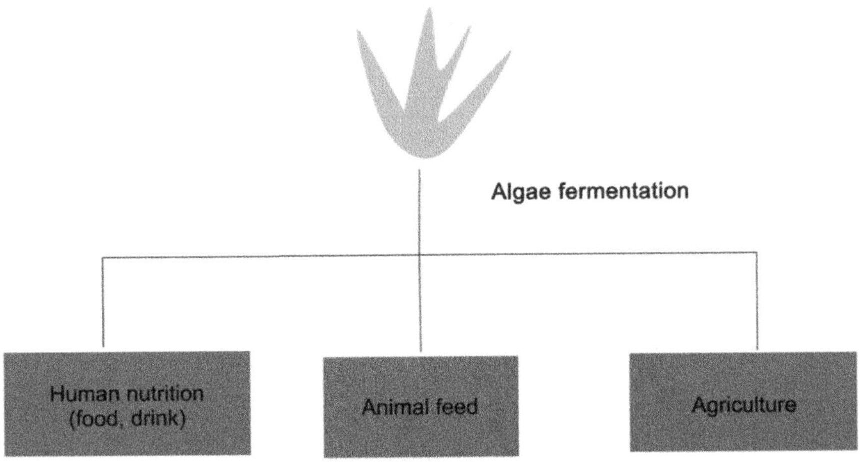

Figure 3.43. *Agricultural and food sectors involved in the development of fermentation processes applied to macroalgae (according to Monteiro et al. (2021))*

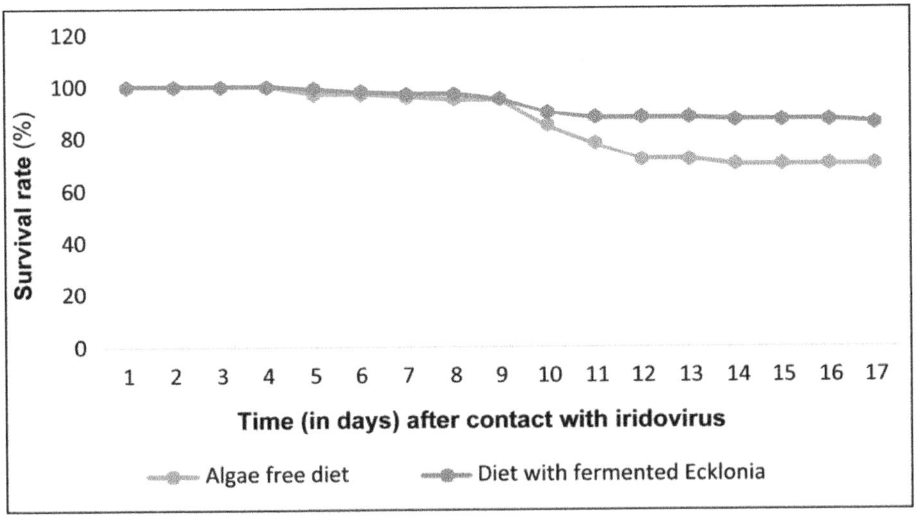

Figure 3.44. *Effect of the addition of fermented algae (10% Ecklonia) on red sea bream survival rate (from Uchida and Miyoshi (2013))*

Algal oligosaccharides obtained by enzymatic or fermentative means are also of interest in terms of plant health. These are derived from the degradation of algal polysaccharides (alginates, laminarin, fucoidans), which are known to induce plant defense mechanisms against pathogens (Vera et al. 2011; Fleurence 2021b). Obtaining oligosaccharides by fermentative means, with a view to developing products for plant health, currently appears to be an interesting avenue for valorizing algae and the fermentation process applied to them.

In addition to the development of new food products (lacto-fermented algae, fermented seaweed juices), fermentation can also be used as a processing method to modify the nutritional value or organoleptic quality of "vegetable seaweed".

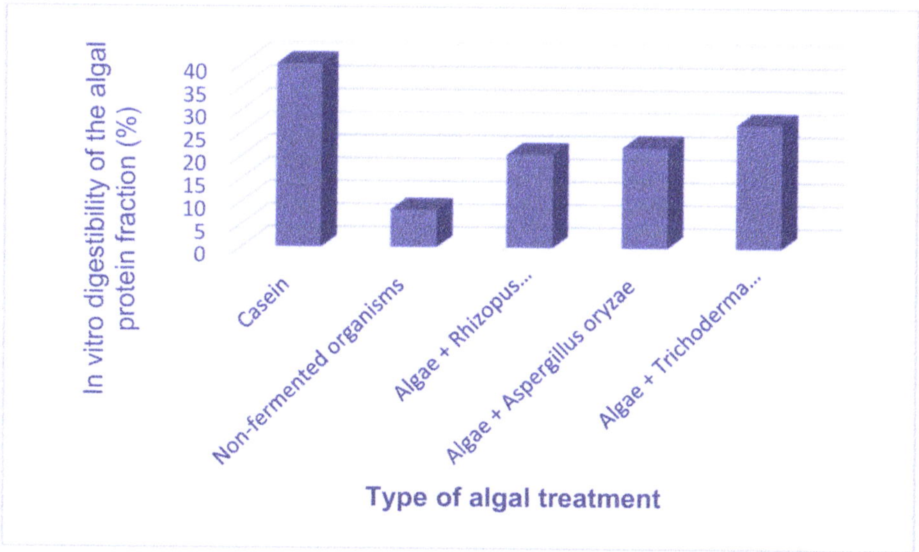

Figure 3.45. *Effect of fermentation of* Palmaria palmata *algae on in vitro protein digestibility after 6 hours (casein: reference protein) (from Marrion et al. (2003))*

The digestibility of seaweed proteins may be limited by the presence of algal polysaccharides such as xylans, which act as nutritional anti-factors (see section 3.1.2). This is particularly true of the seaweed *Palmaria palmata* (Dulse), consumed as a sea vegetable in Europe. In this species, the use of fermentative fungi significantly improves the digestibility of the algal protein fraction compared with that of unfermented seaweed (see

Figure 3.45) (Marrion et al. 2003). Fermentation increases the digestion efficiency of proteins by a minimum factor of 2.5 and a maximum of 3.3. The maximum increase is observed for fermentation via *Trichoderma pseudokoningii*. This fungus produces xylanases in particular, which explains its greater efficiency compared with the other fungal strains tested. In particular, it reduces the concentration of xylans, soluble fibers that inhibit digestion of the algal protein fraction (see Figure 4.46). This last result confirms that soluble xylans are partial inhibitors of protein digestibility. It also shows that pre-treatment of the algae via exogenous xylanases (see section 3.1.2) or via fermentation by a xylanase-producing organism are suitable biotechnological approaches for improving the nutritional value of *P. palmata* proteins.

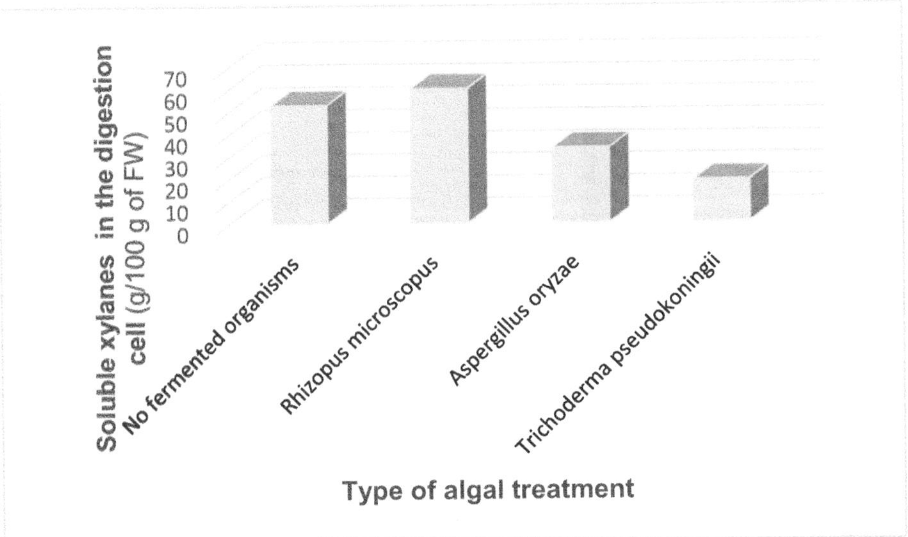

Figure 3.46. *Effect of the fermentation process on the soluble xylan content in the in vitro digestion medium of the algal protein fraction (from Marrion et al. (2003))*

Aspergillus oryzae is a fungus frequently used in Japan, Taiwan and Malaysia to make miso or soy sauce. It is less efficient than *T. pseudokoningii* in improving protein digestibility and degrading soluble fibers. However, its traditional use in soybean fermentation could prove an asset for an industrial application integrating the fermentative transformation of raw materials of different origins (soybean, algae).

Maturation of algae is considered to be an endogenous fermentative process. It effectively relies on the activity of micro-organisms present inside the seaweed or on its surface. The impact of this process on the organoleptical qualities (odors, flavors, texture) of the matured seaweed is a decisive criterion for valorization in the food industry. Organoleptic criteria are assessed by a sensory analysis panel using a rating scale ranging from 1 to 100. This method is used in particular to evaluate the effect of maturation on semi-dehydrated seaweed (20% water) and on dehydrated seaweed (6% water). Maturation is achieved by storing the algae in the dark at 12°C for 126 days. This process applied to *P. palmata* algae shows that maturation has a different impact depending on the moisture content of the starting material. The aroma profile and texture of matured products vary according to this criterion (Stévant et al. 2020; Fleurence 2022). The smell of seaweed or fish skin is more pronounced with the semi-dehydrated product (see Figure 3.47). Preservation of the marine aroma during ripening is therefore better with semi-dehydrated seaweed.

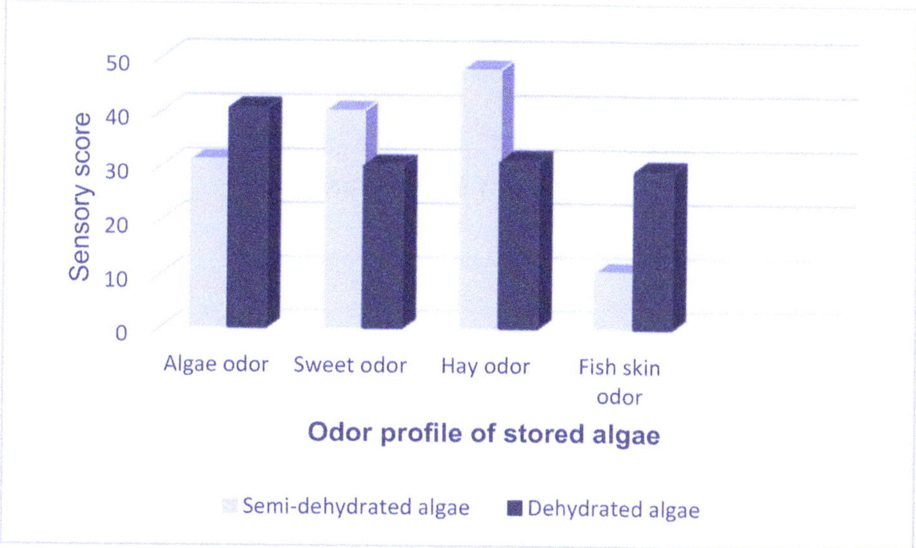

Figure 3.47. *Effect of ripening on the odor profile of* Palmaria palmata *in semi-dehydrated or dehydrated form during storage for 126 days at 12°C (from Stévant et al. 2020)). For a color version of this figure, see www.iste.co.uk/fleurence/aquatic.zip*

As far as texture is concerned, the moisture content of the starting product also determines this criterion (see Figure 3.48). The texture of semi-dehydrated seaweed after 126 days of maturation is softer than that of dehydrated seaweed. Similarly, the crunchiness of semi-dehydrated seaweed is less pronounced than that of dehydrated seaweed.

Seaweed maturation is an interesting process for enhancing a product's organoleptic qualities. This endogenous fermentative process, which relies on storage, is less costly than a process involving the addition of exogenous fermentative organisms. It is therefore ideal for the development of low-value-added fermented products.

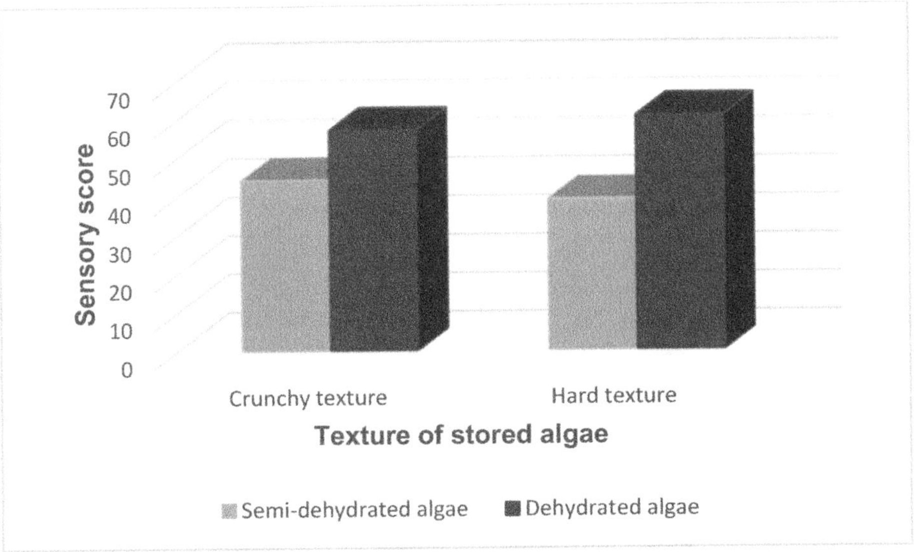

Figure 3.48. *Effect of ripening on the texture of* Palmaria palmata *in semi-dehydrated or dehydrated form during storage for 126 days at 12°C (from Stévant et al. (2020)). For a color version of this figure, see www.iste.co.uk/fleurence/aquatic.zip*

Biotechnological processes (enzymatic engineering, fermentation) applied to the transformation of marine resources make a major contribution to the value chain of the aquaculture and fishing industries. They contribute to the valorization of biomass and its by-products in areas as varied as food, animal feed and fine chemicals.

In all these respects, like genetic engineering, they form part of the current concept of blue biotechnologies.

4

Products and Markets

Blue biotechnologies have given rise to numerous products, some of which remain objects of study or even models for fundamental or applied research. In addition to these, there are other products on the market, generally aimed at the agri-food sector.

4.1. Some examples of marketed and traditional products

In the field of transgenic fish, only AquAdvantage salmon is marketed, on a fairly limited basis and only on the American continent (Canada, USA). AquaBounty aims to produce 55,000 tons per year by 2028 (see section 2.1.2). For the time being, the other genetically modified species form part of aquaculture research and development projects and are not officially marketed as a food product.

Fermentation is a traditional method of preserving fish and has given rise to a large number of marketed products and culinary preparations. These include *surströmming*, a Swedish fish-based speciality (see section 3.2.1). It consists of herring fermented in cans for several months. The smell generated by fermentation is very strong and often surprises the uninformed consumer.

Southern Europe also offers fermented fish products. These products are derived from an endogenous maturation or fermentation process or a fermentation process known as *anchoitage* (see section 3.2.1). In the Mediterranean region, this process is applied to the processing of European anchovies (*Engraulis encrasicolus*). This leads to the production of semi-preserved anchovies, the most common form of anchovy marketed in Europe.

Figure 4.1. *Example of a fermented herring product (surströmming) (photo credit © Flickr, Wikimedia/CC BY 2.0)*

In Asia, Thai fermented fish are a traditional product from North Vietnam. This product is made from freshwater species such as common carp (*Cyprinus carpio*) or crucian carp (*Carassius carassius*), also known as moon carp. The fermentation process takes place in a jar. A mixture of rice powder and sugar is placed at the bottom of the jar, with the fish fillets piled in layers on top. This last condition is essential to obtain homogeneous fermentation. Fermentation lasts a minimum of one week and a maximum of several months.

In Japan, fermentation is also a traditional process for conserving and processing fish. Nare-suzhi or narezushi is a traditional product consisting of gutted, salted fish coated with a layer of fermented rice (see Figure 4.2). This method of preparation enables the fish to be preserved over several months. The fermented fish, stripped of its rice coating, is then consumed as it is. This product is still enjoyed in its native Nara and also in many Japanese restaurants around the world.

In Shiga prefecture, a traditional fermented fish product is also offered to consumers. This is funazushi (see Figure 4.3). This product is made from crucian carp, which is brined and preserved in fermenting rice. The resulting fermented preparation is then eaten as a traditional dish.

Figure 4.2. *Nare-suzhi (photo credit © おむこさん志望 – おむこさん志望, Wikimedia/CC BY-SA 3.0)*

Figure 4.3. *Fermented fish dish (funazushi) (photo credit © WikiTaro 2022, Wikimedia/public domain)*

The bacteria involved in the fermentation of this traditional product have now been identified. Whether in homemade products or in marketed products, they are lactic acid bacteria (Tsuda et al. 2012). However, the flora concerned differs according to the origin of the product (see Table 4.1). In homemade products, the lactic flora is more varied, with *L. plantarum* and *L. buchneri* strains predominating. In contrast, diversity is lower in commercialized products, i.e. those of industrial origin.

Commercial product	Homemade product
Streptococcus salivarius	*Lactobacillus plantarum*
Lactobacillus buchneri	*Lactobacillus buchneri*
Lactobacillus parabuchneri	*Lactobacillus alimentarius*
–	*Lactobacillus farciminis*
–	*Lactobacillus acidipiscis*
–	*Lactobacillus casei*

Table 4.1. *Distribution of lactic acid bacterial floras in funazushi fermented product according to product origin (homemade or industrial) (from Tsuda et al. (2012))*

In Africa, fish fermentation is a traditional process used for stabilizing and preserving this food resource. For most products, fermentation is carried out using fish that have been gutted and salted beforehand (guedj, adjuevan, lanhoui) (see section 3.2.1). There are also fermented products that are further processed by smoking. This is the case for kong, made from catfish or marine catfish (*Arius heudelotii)*, where spontaneous fermentation occurs during artisanal drying prior to the smoking process (see Figure 4.4).

Figure 4.4. *Traditional smoking of catfish (kong) in Senegal (photo credit © Gol T., Cirad 2010)*

Kong is intended for the local market (Senegal) and also for export. In the latter case, product preservation is a major issue. Spraying a culture of *Lactobacillus plantarum* on the smoked product significantly improves (+6 days) the preservation of the product's organoleptic quality at 4°C (African Food Tradition Revisited by Research 2014).

Fermented fish products include sauce. This type of fermented product originates from the Asian culinary tradition. This fermented sauce is known by several names depending on its geographical origin (see Table 4.2). Its Vietnamese name, nuoc mam, has rapidly become a universal reference for this product.

Geographical origin	Name
China (Cantonese)	Jyu lou
China (Mandarin)	Yùlù
Indonesia	Kecap ikan
Taiwan (Minnan)	Hê-iû
Cambodia (Khmer)	Teuk trey
Thailand (Thai)	Nam pla
Vietnam	Nuoc mam
Philippines	Patis

Table 4.2. *Names for traditional fermented fish sauce by geographical origin (from Wikipedia)*

Nuoc mam from the Phu Quoc region of Vietnam is the best known and since 2013 has benefited from a protected designation of origin in Europe. It is made by fermenting anchovies in salt. The layers of fish and salt are alternated, and the whole is left to mature in wooden vats for between 6 and 12 months. The amber juice from fermentation is then collected.

Nuoc mam, a traditional Asian product, seems to have had predecessors in European antiquity. Indeed, a sauce made by fermenting fish flesh in high concentrations of salt was used in Rome and ancient Greece to garnish dishes. This sauce was known as garum (juice) or liquamen (sauce) and may have been similar to today's nuoc mam.

Fermentation, and more specifically lacto-fermentation, can also be applied to seaweed as a processing method. However, the number of products involved is limited compared to fish-based products. France, and in particular Brittany, is one of the pioneers in the production of lacto-fermented seaweed and derived products. The species mainly involved in the fermentative approach is the brown seaweed *Himanthalia elongata*, known under the common name of thongweed or sea spaghetti (see Figure 4.5).

Figure 4.5. Himanthalia elongata *or sea spaghetti used as raw material for lacto-fermentation (photo credit © Mahé S., Algroupe 2023)*

Lacto-fermentation is carried out in buckets using a process and know-how that remain confidential (see Figure 4.6). This process produces very different products.

The first type of product, used as a "sea vegetable", is lacto-fermented seaweed packed in glass jars (see Figure 4.7).

The second type of product generated by the lacto-fermentation of seaweed is a juice that can be used as an ingredient by the food industry (see Figure 4.8).

In addition to *H. elongata*, another brown seaweed, *Saccharina latissima*, formerly *Laminaria saccharina*, is also offered as a lacto-fermented product. Like fermented sea beans, this product is marketed to the general public and caterers under the name royal kombu.

Figure 4.6. Himanthalia elongata *during lactic acid fermentation in storage buckets (photo credit © Mahé S., Algroupe 2023)*

Figure 4.7. *Fermented* Himanthalia elongata *packaged in jars for use as a "sea vegetable" (photo credit © Mahé S., Algroupe 2023)*

Vegetable fermentation is a traditional process in European countries. The products derived from it are well-known and well-accepted by consumers.

The development of the seaweed market is likely to facilitate the growth of fermented products.

Figure 4.8. *Fermented* Himanthalia elongata *juice for use as an ingredient in food (photo credit © Mahé S., Algroupe 2023)*

4.2. Main markets

The market for the production of transgenic fish for food purposes is currently limited to North America (USA, Canada). Information on tons sold is provided by the main player in the industry and lacks external validation (see sections 2.1.2 and 4.1). Similarly, information concerning customers in their main market (Canada) is totally confidential.

The market for fermented fish products is more traditional and therefore has better information. However, this market is highly segmented by country, or even by region, and is difficult to assess globally.

In Africa, and Senegal in particular, the market for guedj (dried fermented fish) (see section 3.2.1) is relatively well-described (Fall et al. 2014). Initially consumed by rural households, guedj was used as a substitute for fresh fish by 93% of households surveyed in the 2014 socio-economic study by Fall et al. Various marine and freshwater species can be fermented to produce guedj. Their frequency of use reflects the importance of the different products in terms of marketing (see Table 4.3) and thus indicates the different market segments involved.

Species (common name)	Frequency of use for making guedj
Catfish	15%
Emperor	15%
False emperor	13%
Meager	12%
Sompat	11%
Pike	10%
Lagna-lagna	10%
Blackspot sea bream	6%
Shark	5%
Black sea bream	3%

Table 4.3. *Frequency of use of different fish species for the production of guedj products in Senegal (from Fall et al. (2014))*

This traditional and popular product is mainly sold on the national Senegalese market. This market is managed by local distributors known as bana-banas. The main guedj-producing regions are Thiès and Ziguinchor. Between them, they account for nearly 76% of guedj production in Senegal (see Figure 4.9).

Most of the guedj produced in Senegal is destined for the domestic market (84%). However, some 16% is exported to other African countries. In this export context, Mali is the main importer of guedj, followed by Guinea-Conakry (see Figure 4.10).

Production of guedj over the period 2006–2015 shows little variation (see Figure 4.11). This implies relative stagnation in this food market. The qualitative evolution of products in terms of biogenic amines (e.g. histamine) and undesirable bacteria is holding back exports of guedj outside Africa. Europe, with its very strict regulations on the quality of processed seafood products, is not an easily accessible market for this type of product.

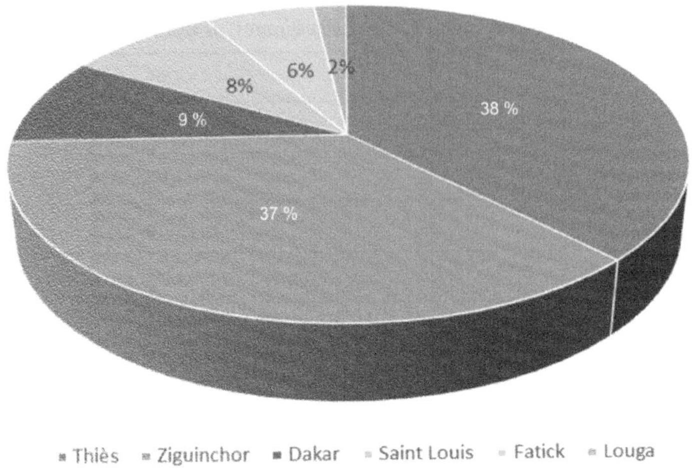

Figure 4.9. *Distribution of guedj production in Senegal by region (from Fall et al. (2014)). For a color version of this figure, see www.iste.co.uk/fleurence/aquatic.zip*

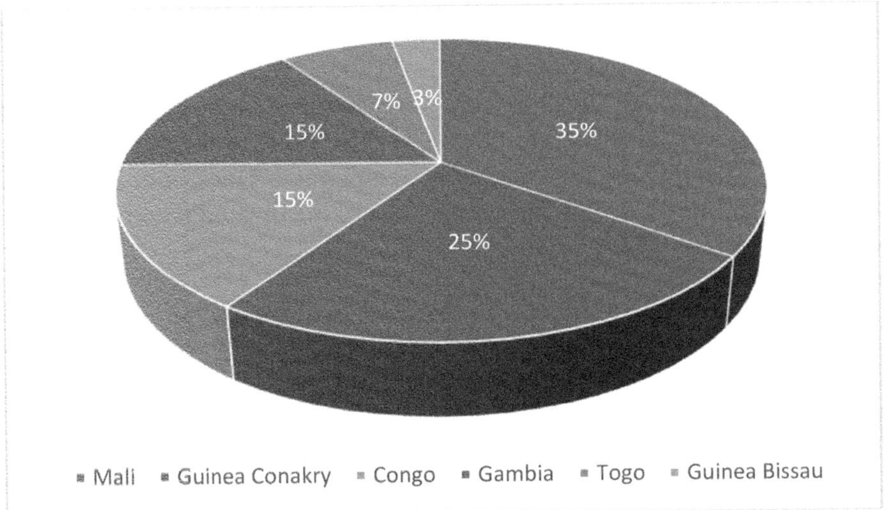

Figure 4.10. *Distribution of guedj production by importing African country (from Fall et al. (2014)). For a color version of this figure, see www.iste.co.uk/fleurence/aquatic.zip*

Despite this constraint, there is a confidential market for fermented fish such as guedj or adjuevan in Europe. This market relies almost exclusively on community or even family trade networks and mainly encompasses expatriates.

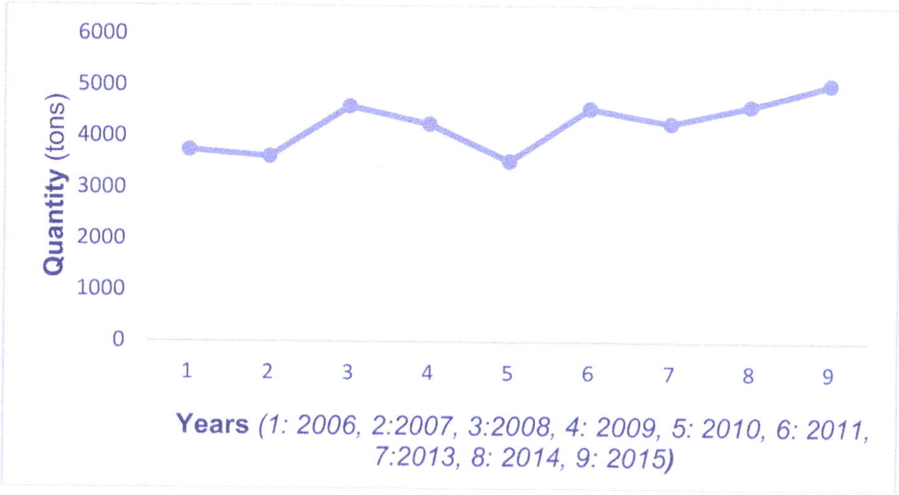

Figure 4.11. *Evolution of annual guedj production in Senegal over the period 2006–2015 (from Fall et al. (2014) and Direction des pêches maritimes du Sénégal (http://www.dpm.gouv.sn))*

Adjuevan is produced mainly in the Ivory Coast. Locally, it is sold on the streets or in markets. Like guedj, adjuevan is made from different species of freshwater or saltwater fish. Many species are common in both guedj and adjuevan, such as lagba-lagba (*Chloroscombrus chrysurus*).

Nationwide, adjuevan is consumed by all population groups (urban, rural), all social classes and all ethnic groups (Kouakou et al. 2013). However, the Akan ethnic group is known to be the biggest consumer of this type of product. Indeed, members of this group consume adjuevan in almost every dish.

The adjuevan production sector in the Ivory Coast is characterized by its informal nature. The fragmentary nature of the information available means that estimates have to be made, particularly in terms of production and markets. In this context, adjuevan production in the Ivory Coast is estimated at between 3,500 and 5,000 tons a year (Kouakou et al. 2013). Just under half of this production is destined for export to third countries such as

Ghana, Burkina Faso, Mali or Benin. In Europe, the product is available from African traders and, in the United States, in supermarkets specializing in exotic products.

In Europe, surströmming is aimed at a very localized market in the Nordic countries and more specifically in Sweden. Although emblematic, this product is not clearly identified in European or FAO statistics. Information on volumes produced and sales generated is therefore national. However, this information is not readily available and is therefore difficult to discuss.

Anchoitage, which leads to the production of semi-preserved anchovies, is a process that was developed in Europe and also in other countries that collect this type of fish (Argentina, Chile). The socio-economic impact of this process and the resulting products is now better understood.

Worldwide, semi-preserved anchovies account for 29% of the 137,300 tons of processed anchovy products (Ababouch and Marrakchi 2009). The main markets for semi-preserved anchovies are Italy, France, Japan, Spain and the USA (see Figure 4.12).

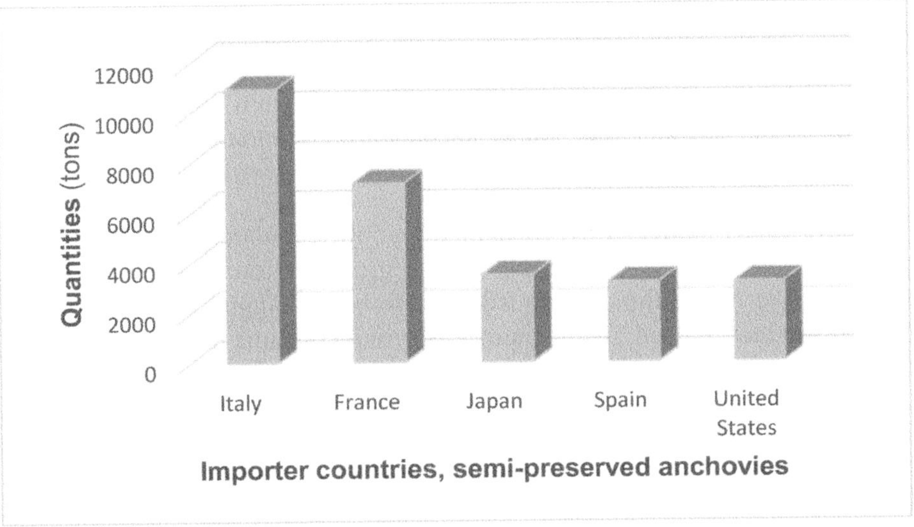

Figure 4.12. *Main markets for semi-preserved anchovies* (anchoitage) *(from Ababouch and Marrakchi (2009))*

Nuoc mam, originally destined for local Vietnamese markets, is now primarily destined for international markets. Vietnam is the world's leading producer, with an annual volume of 170–180 million liters of fish sauce. The domestic market absorbs 30% of this production, with the remainder destined for export.

At present, there are no algal products processed by enzymatic hydrolysis for human or animal consumption. The cost of processing is incompatible with the food market, which generates a low economic margin. Moreover, depending on the enzymes used, regulatory constraints may apply, limiting the marketing authorization of processed products (see Chapter 5). Lacto-fermented algae are not affected by the two aspects mentioned previously. Indeed, fermentation is a process classically used in the food industry, and there is a market for this type of product. Despite this, the market for lacto-fermented algae is limited. In fact, they are mainly marketed to people who are aware of the benefits of seaweed or to restaurateurs. The volumes traded and the sales generated are included without distinction in statistics for the seaweed food industry. France, the pioneer of vegetable seaweed in Europe, is the main producer and market for lacto-fermented seaweed, although it is difficult to assess its relative importance compared to other seaweed-based food products (dehydrated, salted, frozen, appertized).

Blue biotechnology products have very different applications in human and animal nutrition. In terms of animal feed, many species (algae, fish) produced by aquaculture or marine harvesting are used in the value chain for livestock nutrition, whether terrestrial or aquatic in origin. The same applies to human nutrition.

Genetically modified organisms, and more specifically AquAdvantage salmon, are beginning to be introduced as a new food product the American market, which is more receptive and favorable than the European market.

Fermentation, an initially empirical and traditional biotech process, generates numerous products from fish or algae matrices. This approach is probably the easiest and most economical to implement.

5

Regulations

Regulation is a legal constraint that applies to all sectors of economic activity. Products derived from blue biotechnologies and the processes used in their manufacture are no exception to this rule. Regulations differ from product to product and from country to country. In some cases, there is a lack of legal clarity or even no regulations at all.

5.1. Transgenic products

Worldwide, few transgenic animals are approved for commercialization. However, some genetically modified fish, such as the GloFish zebrafish, may be marketed for aquarium purposes. This fish is modified to produce a green fluorescent protein from a transgene derived from a jellyfish. The sale or possession of GloFish not legally purchased is prohibited in the United States. In California, this prohibition also applies to GloFish legally purchased in another state.

Among genetically modified fish, AquAdvantage salmon is the only one to be produced in aquaculture for human consumption. The FDA approved its marketing as a food in 2015. The preamble to the approval by the federal administration is as follows: "The salmon involved is safe to consume and the DNA introduced is safe for the fish itself."[1]

In this decision, the health risk assessment is based primarily on the study of nutritional equivalence (see section 2.1.2). This equivalence is based on a

1 See: https://www.fda.gov/animal-veterinary/aquadvantage-salmon/aquadvantage-salmon-fact-sheet.

comparison between the two fish in terms of the main biochemical constituents of muscle (proteins, carbohydrates, lipids), their water content and their mineral composition. On this basis, the FDA considers that the nutritional profile of transgenic salmon, and therefore the quality of the resulting food, is comparable to that of non-transgenic salmon raised by aquaculture. Similarly, a study carried out on the hormonal profile shows no significant differences between the two types of fish. The risk of hormonal overdosage following consumption of AquAdvantage salmon is therefore considered negligible by the authorities.

Irrespective of health aspects, the FDA recommends that AquAdvantage salmon be labeled to specify the transgenic origin when marketed. In the United States, this regulatory requirement applies to all genetically modified products intended for human consumption.

As such, transgenic salmon falls under the National Bioengineered Food Disclosure Act, enforced by the State Department of Agriculture[2].

In addition to regulatory considerations relating to the product, the FDA has also established constraints on the production of transgenic salmon. These conditions are strict. To avoid any risk of escape, AquAdvantage salmon cannot be raised in the wild in pens with ocean nets, as is the case for non-transgenic salmon. They must be produced only in specific land-based facilities. Production in these facilities is limited to two geographical sites: Canada, where the transgenic broodstock and egg stock are kept, and the state of Indiana (Albany), where the market fish are to be raised.

This FDA opinion has sometimes been contested at state level. Numerous bills aimed at banning it have been introduced, and some have been passed at the local level. On September 19, 2014, the Governor of California signed a bill (AB504) banning the production of transgenic salmonids from all California waters. This ban covers AquAdvantage salmon, of course, and also fario trout, another target salmonid for AquaBounty Technologies.

In the European Union, no transgenic animals are authorized for sale[3]. This effectively restricts the European market to AquAdvantage salmon and GloFish zebrafish.

---

2 See: https://www.ams.usda.gov/rules-regulations/be.

3 See: www.ecologie.gouv.fr/organismes-genetiquement-modifies-ogm-0.

5.2. Other products

Aquatic products, whether of marine or freshwater origin, are subject in Europe and particularly in France to regulations governing traceability (name, origin).

In 2022, the French Directorate General for Competition, Consumer Affairs and Fraud Prevention (DGCCRF) drew up a list of names accepted in France for the marketing of fish and aquaculture products[4]. This list is an application of a European regulation (European Union 2013b), which defines the traceability and labeling rules applicable to fishery and aquaculture products. As the DGCCRF points out, this list in no way prejudges authorization for commercialization of the species on it in France.

Regulations may also define toxicological criteria to protect consumer health. The most closely followed criterion concerns histamine content. This biogenic amine is responsible for sometimes severe food poisoning (anaphylactic shock), the symptoms of which can be confused with those of food allergy. This amine is produced by the breakdown of histidine during the post-mortem alteration of fish flesh, and its content increases as the product's freshness decreases. A rise in histamine levels can also be observed during fish maturation or fermentation processes. Fermented fish sauce is a particularly sensitive product in this respect. Epidemiological studies linking histamine content with the risk of toxicity to the consumer have made it possible to refine regulations on the subject (see Table 5.1).

Histamine content (mg/kg)	Toxicity risk level
<50	No risk
50–200	Possible risk
200–1,000	Probable risk
>1,000	Certain risk

Table 5.1. *Assessment of intoxication risk according to histamine content in fish and fish-derived products (according to ANSES (2012))*

4 See: https://www.economie.gouv.fr/dgccrf/Consommation/Etiquetage-des-produits/Produits-de-la-mer-et-d-eau-douce/Denominations-commerciales-des-produits-de-la-peche-et-de-l-aqua culture.

On the basis of available studies, the European Union has defined histamine levels that must not be exceeded in different food categories (European Union 2013a).

For processed foods from histidine-rich species, the maximum histamine content is 200 mg/kg of product. For products derived from enzymatic maturation or fermentation in brine, the histamine content must not exceed 400 mg/kg of product. As before, this limit applies to species rich in histidine. These include herring and anchovy, species frequently used in matured or fermented fish products.

By way of comparison, the FDA in the USA sets a limit of 500 mg/kg histamine in products from histamine-rich fish species.

In addition to the regulations applied to matured or fermented fish products, seaweed, whether harvested or cultivated, is subject to very different regulations in different countries.

Indeed, the use of seaweed as a foodstuff may be subject to very specific regulations or no regulations at all. In Europe, France is the only country to have introduced restrictive regulations on the use of seaweed for human consumption. These specific regulations apply to all derived products, including enzymatically processed, matured or fermented algae. Historically, seaweed has been classified by the DGCCRF as a "non-traditional food". Its marketing as a foodstuff was subject to the prior approval of the Superior Council of Public Hygiene of France and the French National Academy of Medicine. A positive opinion was given in 1990, covering 11 species of macroalgae and 1 species of cyanobacteria (Fleurence 1991; Mabeau and Fleurence 1993). Today, the list of algae authorized for human consumption has been considerably extended (see Table 5.2) and now includes 22 macroalgae, 2 microalgae and 1 cyanobacterium.

Only the algae listed here may be marketed in France, regardless of the treatment applied (dehydration, freezing, appertization, enzymatic maceration, fermentation, etc.).

Products derived from the processing of these species, or algae with little or no processing, must also meet toxicological and bacteriological criteria. Toxicological criteria concern heavy metals and iodine (see Table 5.3). They apply to all seaweed-based products, including those derived from enzymatic or fermentative maturation processes.

	Species	Common name
Red algae	*Palmaria palmata*	Dulse
	Porphyra umbilicalis	Nori
	Porphyra tenera	Nori
	Porphyra yezoensis	Nori
	Porphyra dioica	Nori
	Porphyra purpurea	Nori
	Porphyra laciniata	Nori
	Porphyra leucostica	Nori
	Chondrus crispus	Pioca, lichen
	Gracilaria verrucosa	Ogonori
	Lithothamnium calcareum	Maerl
Brown algae	*Ascophyllum nodosum*	Seaweed
	Fucus serratus	Kelp
	Fucus vesiculosus	Black sea bream
	Himanthalia elongata	Sea spaghetti
	Laminaria digitata	Breton kombu
	Laminaria saccharina	Royal kombu
	Laminaria japonica	Kombu
	Undaria pinnatifida	Wakame
	Alaria esculanta	Atlantic wakame
Green algae	*Ulva* sp.	Sea lettuce
	Enteromorpha sp.	Aonori
Microalgae	*Odontella aurita*	–
	Chorella sp.	Chlorella
	Spirulina sp.	Spirulina

Table 5.2. *Updated list of algae (macro- and microalgae) and cyanobacteria authorized in France for use as sea vegetables, intermediate food products or ingredients (according to Fleurence (2018))*

Bacteriological criteria are associated with the determination of aerobic and anaerobic floras, as well as the search for pathogenic germs such as salmonella or staphylococcus (see Table 5.4). Bacteriological criteria were originally established for the main category of products marketed: dehydrated seaweed. Since then, other derived products such as lacto-fermented seaweed algae have appeared. During processing, these seaweeds are inoculated with lactic acid bacteria to increase the level of anaerobic flora. Certain bacteriological criteria, initially established for dried seaweed, are therefore ill-suited to seaweed processed by fermentation.

Elements	Maximum content (mg/kg DW)
Mineral arsenic (inorganic)	≤ 3
Cadmium	≤ 0.5
Pewter	≤ 5
Lead	≤ 5
Mercury	≤ 0.1
Iodine	$\leq 2,000$

Table 5.3. *Maximum levels of heavy metals and iodine permissible in edible seaweed and seaweed products (from Fleurence (2018))*

Germs	Quantity (germs/g DW)
Mesophilic aerobic flora	$\leq 100,000$
Fecal coliforms	≤ 10
Anaerobic sulfite-reducing flora	≤ 100
Staphylococcus aureus	≤ 10
Clostridium perfringens	≤ 1
Salmonella sp.	Absence in 25 g

Table 5.4. *Microbiological criteria applied to dehydrated seaweed (according to Fleurence (2018))*

European regulations are more permissive than French ones. The EU has not established a positive list of species authorized as food ingredients. Instead, seaweed is included in a European Parliament and Council regulation on novel foods and novel food ingredients (Reg. EC no. 258/97). This regulation, known as "Novel Foods", stipulates in Article 2, paragraph d, that:

> This Regulation shall apply to the placing on the market in the Community of foods and food ingredients for which human consumption has so far remained negligible in the Community and which fall into the following categories: [...] d) foods and ingredients composed of or isolated from micro-organisms, fungi or algae.

The regulation also authorizes the marketing of foods or food ingredients containing or derived from genetically modified organisms, but which do not contain a transgene in the final product. This last point does not apply to transgenic salmon, the production and commercialization of which are banned in Europe. Should the production ban be lifted, transgenic salmon would have to meet the requirements of Article 10 of the European Council Directive defining the conditions for the marketing of GMOs (European Community 1990).

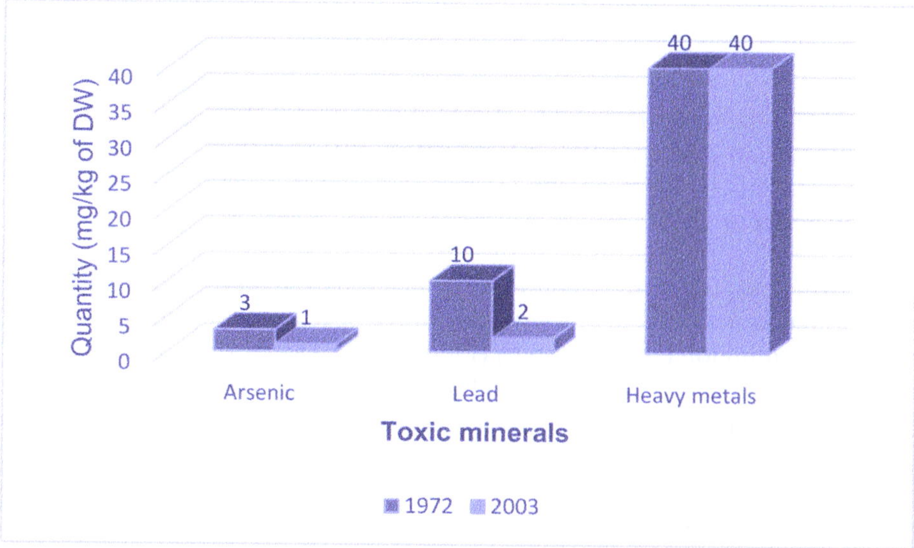

Figure 5.1. *Evolution of acceptability levels for arsenic, lead and heavy metals in food algae according to the Food Chemical Codex between 1972 and 2003 (from Fleurence (2018)). For a color version of this figure, see www.iste.co.uk/fleurence/aquatic.zip*

In the United States, seaweed intended for human consumption must meet quality and health criteria laid down in the Food Chemical Codex. Historically (1972), these criteria were defined for the main species harvested in the country, namely *Macrocystis pyrifera*, more commonly known as Pacific kelp. They concern dry matter, ash content and, of course, the content of certain toxic minerals such as arsenic, lead and iodine (Fleurence 2018). In 2003, the Food Chemical Codex established new values, notably lowering the maximum acceptable limit for arsenic and lead (see Figure 5.1). Between the two dates, there was a notable change in the recommended iodine content. Unlike heavy metals, its level of acceptability is clearly higher in the recommendations published in 2003 (see Table 5.5). The same applies to ash content.

	Food Chemical Codex (1972)	Food Chemical Codex (2003)
Dry matter	≤ 13% FW	≤ 13% FW
Ash	≤ 35% DW	≤ 45% DW
Iodine	0.15–0.22% DW	0.1–0.5% DW

Table 5.5. *Evolution of quality criteria and iodine content in edible algae according to the Food Chemical Codex between 1972 and 2003 (from Fleurence (2018))*

In 2003, the Food Chemical Codex also specified that these criteria apply to *Macrocystis pyrifera* and all brown algae belonging to the genus *Laminaria*. This includes kombu (*Laminaria japonica*), the world's leading food-grade seaweed.

In 2017, the Code of Federal Regulations took the regulatory approach a step further. It defines a list of brown seaweed species recognized as safe for human consumption (see Table 5.6) (Fleurence 2018). This list complements the one initially defined by the Food Chemical Codex. In this new regulation, the use of seaweed is mainly authorized, in dehydrated crushed or chopped form, as an ingredient or intermediate food product. The recommendations on its use as ingredients concern spices, seasonings or flavorings. The regulations applied to brown seaweed also include indications of geographical origin. The seaweed concerned must be harvested in the coastal waters of the North Atlantic and Pacific Oceans.

In the same context, the Code of Federal Regulations establishes a list of red algae that can be used for human consumption in the same forms as above and for the same type of application (see Table 5.7). As in the case of brown seaweed, the regulations give geographical indications of the harvesting areas for authorized species. For species of the genus *Porphyra*, authorized algae must originate from the coasts of Japan, Korea, China or Taiwan. For *Rhodymenia palmata*, the geographical origin is limited to the east and west coasts of the United States. This excludes, ipso facto, authorization of the same species produced on the west European coasts of the Atlantic Ocean. For the *Gloiopeltis furcata* species, the geographical area covered by the regulations mainly concerns the southern coasts of the Pacific Ocean.

Species	Spices	Seasoning	Aromas
Analipus japonicus	x	x	x
Eisenia bicyclis	x	x	x
Hizikia fusiforme	x	x	x
Kjellmaniella gyrata	x	x	x
Laminaria angustata	x	x	x
Laminaria claustonia	x	x	x
Laminaria digitata	x	x	x
Laminaria longicruris	x	x	x
Laminaria longissima	x	x	x
Laminaria ochotensis	x	x	x
Laminaria saccharina	x	x	x
Laminaria japonica	x	x	x
Macrocystis pyrifera	x	x	x
Petalonia fascia	x	x	x
Scytosiphon lomentaria	x	x	x
Undaria pinnatifida	x	x	x

Table 5.6. *Brown seaweed species authorized in the Code of Federal Regulations for human consumption and areas of application (according to the Code of Federal Regulations (2017a) and Fleurence (2018))*

Species	Spices	Seasoning	Aromas
Gloiopeltis furcata	x	x	x
Porphyra crispata	x	x	x
Porphyra dentata	x	x	x
Porphyra perforata	x	x	x
Porphyra suborbiculata	x	x	x
Porphyra tenera	x	x	x
Rhodymenia palmata *(Palmaria palmata)*	x	x	x

Table 5.7. *Red algae species authorized in the Code of Federal Regulations for human consumption and areas of application (according to the Code of Federal Regulations (2017b) and Fleurence (2018))*

American regulations governing seaweed used as human food are ultimately more restrictive than French regulations. Like the latter, they establish a positive list of seaweed species that can be used. They also recommend health criteria and impose limited areas of application for ingredients and harvesting or cultivation zones for authorized species.

Some countries take a different approach to the regulation of seaweed. This is the case with Australia and New Zealand. In these countries, seaweed suitable for consumption is assessed according to food safety standards. Failure to meet these criteria results in the product being classified as unsafe. If imported, the Australian and New Zealand authorities place the product on the list of foodstuffs not considered as safe, and it may be blocked.

The two main criteria taken into account by the Australian and New Zealand authorities are levels of iodine and arsenic. For iodine, the maximum level cannot exceed 1 g iodine/kg dry matter[5]. For inorganic arsenic, the maximum value is set at 1 mg/kg of dry matter. A study carried out by these authorities shows that most edible seaweeds marketed in dehydrated form meet this criterion (see Table 5.8). However, the product hijiki (*Hizikia fusiforme* or *Sargassum fusiforme*) often does not meet this standard. This product is therefore classified as a risk by the authorities, who are continuing to gather data on this seaweed, which is highly prized in Japan and imported into Australia and New Zealand.

Dehydrated seaweed	Inorganic arsenic (mg/kg DW)			
	Australia	New Zealand	Spain	United Kingdom
Wakame	0.16–0.20	0.1	0.15–0.26	< 0.3
Kombu	0.16–0.33	n.d.	0.25–0.30	< 0.3
Hijki	0.32–7.8	n.d.	83–85	77
Nori	0.09–0.16	1.2	0.19–0.57	< 0.3

Table 5.8. *Examples of inorganic arsenic dehydrated seaweed from products marketed in various countries (from Food Standards Australia New Zealand)*[6]

5 See: https://www.foodstandards.gov.au/science/surveillance/Documents/Iodine%20in%20Seaweed.pdf.

6 See: https://www.foodstandards.gov.au/science/surveillance/Documents/Survey%20of%20inorganic%20arsenic%20in%20seaweed%20and%20seaweed%20.pdf.

In addition to these criteria, recommendations are also proposed for the consumption of brown seaweed by pregnant or breastfeeding women. In particular, these seaweeds, which are naturally rich in iodine, should not be consumed more than once a week. A list of species and products concerned is given by the authorities (see Table 5.9).

Species	Trade names	Product types
Laminaria* spp.** ***Laminaria japonica	Kelp, kombu, suboshi kombu, aokombu, haida dashima, oarweed, wild kelp, fingered tangle, sea tangle	Dashi, salad, fried products, soup, sauce, seaweed tea, drinks, sahimi
***Undaria* spp.**	Wakame, haiboshi wakame, quandai-cai, miyeok, sea fern, mekabu	Miso soup, salads
Sargassum fusiforme ***Sargassum* spp.**	Hiziki, hoshi hiziki, deer tail grass, sheep nest grass, horse tail tangle	Vegetables, soup, sautéed fried products
Cladosiphon okamuranus	Mozuku	Salads, food supplements
Alaria esculenta	Winged kelp, dabberlocks, badderlocks, American wakame	Salads, vegetables
Eisenia bicyclis	Arame, kelp	Vegetables, salads, garnishes, pickles
Ecklonia maxima	Sea bamboo, paddle weed	Vegetables, condiments
Fucus vesiculosus	Bladder wrack, black tang, rockweed, bladder fucus, sea oak, black tany, cut weed, dyers fucus, red fucus, rock wrack	Additives, flavorings, food supplements
Durvillaea antarctica	Bull kelp, cochayuyo, hulpe	Stews, salads
Ascophyllum nodosum	Norwegian kelp, knotted kelp, knotted wrack, egg wrack	Alginates, seaweed flour
Postelsia palmaeformis	Sea palm, palm seaweed	Chinese dishes

Table 5.9. *List of brown seaweeds rich in iodine for which consumption restrictions have been set for pregnant and breastfeeding women (according to Food Standards Australia New Zealand)*[7]

7 See: https://www.foodstandards.gov.au/consumer/safety/brownseaweed/pages/brownseaweedstable.aspx.

The Australian and New Zealand regulations establish stricter criteria for maximum values of iodine and inorganic arsenic for edible algae than those defined in the French regulations. However, they do not establish a positive list in terms of species.

Marketed seaweed, mainly imported, must comply with the values set out by Food Standards Australia New Zealand. This database, which is updated over time, is the regulatory platform on which the authorities in these two countries rely[8].

Asian countries (China, Korea, Japan) where seaweed is traditionally consumed have not developed any specific regulations concerning its use in human food. The products marketed simply have to meet standard sanitary criteria for foodstuffs. However, some countries have introduced regulatory constraints that are highly specific to imported seaweed. This is the case in Japan, which imposes specific criteria for heavy metals, applicable only to imported seaweed.

This protectionist action can also be reinforced by regulatory action defining a maximum quota for imported seaweed. This was the case in the early 2000s for imports of dried and seasoned *Porphyra* seaweed. This limitation affected imports from South Korea. It has been the subject of a trade dispute between Korea and Japan, which has led to an appeal to the World Trade Organization by Korea[9].

The regulations applied to blue biotechnology products vary widely, depending on product category and country.

As for genetically modified organisms such as transgenic fish, most countries do not authorize this type of product. Only the FDA (USA) authorizes the production and marketing of AquAdvantage salmon. Canada has given its approval for sale. For products derived from enzymatic transformation, maturation and fermentation processes, there are no regulatory restrictions, unless the enzymes or ferments are not authorized for human food applications. Seaweed, an aquatic resource traditionally

8 See: https://www.foodstandards.gov.au/Pages/default.aspx.

9 See: https://www.wto.org/french/tratop_f/dispu_f/cases_f/ds323_f.htm.

consumed in Asia and to a lesser extent in the West, is treated in very different ways in different countries.

Countries that have traditionally included seaweed in their diet have no specific regulations governing this type of foodstuff. In contrast, countries that are discovering this foodstuff, through exports or the development of a new aquaculture sector, tend to introduce regulations that are more or less permissive but specific to edible algae.

Conclusion

Blue biotechnologies are an important part of the biotechnology industry. They are applied specifically to aquatic resources, whether of marine, freshwater, plant or animal origin. They are based on the controlled or empirical transformation of living organisms for food or other purposes. This biotransformation calls on a variety of disciplinary approaches: developmental biology for the aquaculture genetic engineering for modifying the genome of organisms and enzymatic and fermentative engineering for food processing.

The production of transgenic fish as model animals for the study of vertebrate development or human pathologies is an example of a transformation that could prove useful. In this case, social acceptance seems easier, as we are a long way from transgenic salmon, whose ultimate purpose is the emergence of a new aquaculture production system known as the blue revolution.

The development of algal farming could, in turn, meet a number of ecological challenges. Increasing the surface area dedicated to macroalgae cultivation is one way of increasing the number of CO_2 sinks on the ocean surface. The use of algae as a source of protein for animal feed to replace soy proteins, the cultivation of which is responsible for the deforestation of the Amazon, is a strategic consideration that should be pursued rapidly (Fleurence 2023).

Similarly, the cultivation of microalgae for biofuel production is a biotechnological approach that can meet both ecological and economic challenges (Fleurence 2021a).

Blue biotechnology also reminds us of the great potential and dynamism of a sector of activity with the task of meeting the demand for food proteins for an ever-growing human population. Fish and seaweed farming, as well as processes for transforming these resources to improve their nutritional value or preservation, are the main aspects of blue biotechnologies.

Blue biotechnologies also teach us that technology applied to the living world cannot be summed up as the application of technical processes without questioning the aims and objectives of this approach to transforming biological resources.

References

Ababouch, L. and Marrakchi, E. (2009). Élaboration des semi-conserves d'anchois : Aspects économiques, techniques et hygiéniques. Technical document, FAO.

African Food Tradition Revisited by Research (2014). Characteristics of the processes and the products from kong favorable to the development of the technological flora. Report D3.3.2, 1–5.

Amano, H. and Noda, H. (1990). Proteins of protoplasts from red alga *Porphyra yezoensis*. *Nippon Suisan Gakkaishi*, 56, 1859–1864.

Amano, H. and Noda, H. (1992). Proteins of protoplasts from several seaweeds. *Nippon Suisan Gakkaishi*, 58, 291–299.

ANSES (2012). Histamine. Fiche de description de danger biologique transmissible par les aliments [Online]. Available at: https://www.anses.fr/fr/system/files/MIC2012sa0209Fi.pdf.

Appenroth, K.J., Sree, K.S., Bog, M., Ecker, J., Seeliger, C., Böhm, V., Lorkowski, S., Sommer, K., Vetter, W., Tolzin-Banasch, K. et al. (2018). Nutritional value of the duckweed species of the genus *Wolffia* (Lemnaceae) as human food. *Frontiers in Chemistry*, 6, 483.

Avila-Peltroche, J., Won, B.Y., Cho, T.O. (2022). An improved protocol for protoplast production, culture, and whole plant regeneration of the commercial brown seaweed *Undaria pinnatifida*. *Algal Research*, 67, 10285.

Benfey, T.J. (2001). Use of sterile triploid Atlantic salmon (*Salmo salar* L.) for aquaculture in New Brunswick, Canada. *ICES Journal of Marine Science*, 58, 525–529.

Benfey, T.J. (2016). Effectiveness of triploidy as a management tool for reproductive containment of farmed fish: Atlantic salmon (*Salmo salar*) as a case study. *Reviews in Aquaculture*, 8, 264–282.

Campello, F. (1983). Microflore et anchoitage. *Revue des travaux de l'Institut des pêches maritimes*, 47, 227–236.

Carvan III, M.J., Dalton, T.P., Stuart, G.W., Nebert, D.W. (2000). Transgenic zebrafish as sentinels for aquatic pollution. *Annals of the New York Academy of Sciences*, 9, 133–147.

Chatakondi, N., Lovell, R.T., Duncan, P.L., Hayat, M., Chen, T.T., Powers, D.A., Weete, J.D., Cummins, K., Dunham, R.A. (1995). Body composition of transgenic common carp, *Cyprinus carpio*, containing rainbow trout growth hormone gene. *Aquaculture*, 138, 99–109.

Chaveesuk, R., Smith, J.P., Simpson, B.K. (1994). Production of fish sauce and acceleration of sauce fermentation using proteolytic enzymes. *Journal of Aquatic Food Product Technology*, 2, 59–77.

Chen, C.S. and Shyu, J.F. (1994). Isolation of protoplasts from four species of brown algae. *Botanical Bulletin of Academia Sinica*, 35, 95–104.

Chen, L.M., Craigie, J.S., Xie, Z.K. (1994). Protoplast production from *Porphyra linearis* using a simplified agarase procedure capable of commercial application. *Journal of Applied Phycology*, 6, 35–39.

Chen, Y.C. and Shih, H.C. (2000). Development of protoplasts of *Ulva fasciata* (Chlorophyta) for algal seed stock. *Journal of Phycology*, 36, 608–615.

Cindana Mo'o, F.R., Wilar, G., Devkota, H.P., Wathoni, N. (2020). Ulvan, a polysaccharide from macroalga *Ulva* sp.: A review of chemistry, biological activities and potential for food and biomedical applications. *Applied Sciences*, 10, 5488.

Code of Federal Regulations (2017a). Direct food substances affirmed as generally recognized as safe. Brown algae. Title 21, 31, section 184.1120.

Code of Federal Regulations (2017b). Direct food substances affirmed as generally recognized as safe. Red algae. Title 21, 31, section 184.1121.

Dumay, J., Barthomeuf, C., Bergé, J.P. (2004). How enzymes may be helpful for upgrading fish by-products: Enhancement of fat extraction. *Journal of Aquatic Food Product Technology*, 13, 69–84.

Dumay, J., Clément, N., Morançais, M., Fleurence, J. (2013). Optimization of hydrolysis conditions of *Palmaria palmata* to enhance R-phycoerythrin extraction. *Bioresource Technology*, 131, 21–27.

El-Sayed, A.F., El-Ghobashy, A., Al-Amoudi, M. (1996). Effects of pond depth and water temperature on the growth, mortality and body composition of Nile tilapia, *Oreochromis niloticus* (L.). *Aquaculture Research*, 27, 681–687.

Eumofa (2017). Étude de cas. La truite fraîche portion en Pologne. Structure des prix dans la filière. Eufoma, 1–42 [Online]. Available at: https://www.eumofa.eu/documents/20178/110724/La+truite+fra %C3 %AEche+en+Pologne_FR.pdf.

Eumofa (2021). Étude de cas. La truite portion dans l'UE. Structure des prix dans la chaîne d'approvisionnement. Focus sur l'Allemagne et la Pologne. Eufoma, 1–65 [Online]. Available at: https://www.eumofa.eu/documents/20178/474612/PTAT+Portion+trout+in+PL+DE+IT_FR.pdf/67955025-0ad6-960a-24c2-253dc966eea9?t=1635330903350.

European Community (1990). Directive du Conseil relative à la dissémination volontaire d'organismes génétiquement modifiés dans l'environnement (90/220/CEE). *Journal officiel des communautés européennes*, 117(15).

European Union (2013a). Règlement (UE) n° 1019/2013 de la Commission du 23 octobre 2013 modifiant l'annexe 1 du règlement (CE) n° 2073/2005 en ce qui concerne l'histamine dans les produits de la pêche. *Journal officiel de l'Union européenne*, 228(46).

European Union (2013b). Règlement (UE) n° 1379/2013 du Parlement européen et du Conseil du 11/12/13 portant organisation commune des marchés dans le secteur des produits de la pêche et de l'aquaculture, modifiant les règlements (CE) n° 1184/2006 et (CE) n° 1224/2009 du Conseil et abrogeant le règlement (CE) n° 104/2000 du Conseil. *Journal officiel de l'Union européenne*, 354(1).

Fall, N.G., Tounkara, L.T., Diop, M.B., Thiaw, O.T., Thonart, P. (2014). Étude socio-économique et technologique de la production du poisson fermenté et séché (Guedj) au Sénégal. *International Journal of Biological and Chemical Sciences*, 8, 2523–2538.

Fall, M., Diop, M.B., Montet, D., Maiga, A.S., Guiro, A.T. (2019). Fermentation du poisson en Afrique de l'Ouest et défis sociétaux pour une amélioration qualitative des produits (adjuevan, guedj et lanhouin) : Revue de la littérature. *Cahiers Agricultures*, 28(7), 1–12.

FAO (2020). La situation mondiale des pêches et de l'aquaculture. La durabilité en action. FAO, 1–247.

FAO (2021a). Fisheries & aquaculture – Cultured aquatic species information programme – *Undaria pinnatifida* ((Harvey) Suringar, 1873). FAO.

FAO (2021b). Global seaweeds and microalgae production, 1950-2019. FAO.

FAO (2022a). Résumé. La situation mondiale des pêches et de l'aquaculture. Vers une transformation bleue. FAO.

FAO (2022b). *Salmo salar*. Programme d'information sur les espèces aquatiques cultivées [Online]. Available at: https://www.fao.org/fishery/fr/culturedspecies/salmo_salar_fr.

Fletcher, G.L. and Davies, P.L. (1991). Transgenic fish for aquaculture. *Genetic Engineering*, 13, 331–370.

Fletcher, G.L., Kao, M.H., Fourney, R.M. (1986). Antifreeze peptides confer freezing resistance to fish. *Canadian Journal of Zoology*, 64(9), 1897–1901.

Fleurence, J. (1991). L'habilitation des algues en alimentation humaine: Le point sur la réglementation française. *Industries alimentaires et agricoles*, 108(6), 501–502.

Fleurence, J. (1999a). The enzymatic degradation of algal cell walls: A useful approach for improving protein accessibility? *Journal of Applied Phycology*, 11, 313–314.

Fleurence, J. (1999b). Seaweed proteins: Biochemical, nutritional aspects and potential uses. *Trends in Food Science and Technology*, 10, 25–28.

Fleurence, J. (2018). *Les algues alimentaires : Bilan et perspectives*. Lavoisier Tec & Doc, Paris.

Fleurence, J. (2021a). *Microalgae: From Future Food to Cellular Factory*. ISTE Ltd, London, and John Wiley & Sons, New York.

Fleurence, J. (2021b). Perspectives on the use of algae in agriculture and animal production. *Phycology*, 1, 79–82.

Fleurence, J. (2022). Biotechnological processes applied to edible seaweeds: What perspectives? *Trends in Food Science and Technology*, 129, 617–620.

Fleurence, J. (2023). *Algae in Agrobiology: Realities and Perspectives*. ISTE Ltd, London, and John Wiley & Sons, New York.

Fleurence, J. and Joubert, Y. (2006). Procédé d'extraction d'ADN incluant une étape de digestion enzymatique. International Patent (PCT), WO 2006/016023 A1.

Fleurence, J., Massiani, L., Guyader, O., Mabeau, S. (1995). Use of enzymatic cell wall degradation for improvement of protein extraction from *Chondrus crispus*, *Gracilaria verrucosa* and *Palmaria palmata*. *Journal of Applied Phycology*, 7, 393–397.

Fleurence, J., Antoine, E., Luçon, M. (2002). Procédé d'extraction et d'amélioration de la digestibilité des protéines de *Palmaria palmata*. International Patent (PCT), WO 02/07528 A1.

Fu, C., Cui, Y., Hung, S.S.O., Zhu, Z. (1998). Growth and feed utilization by F4 human growth hormone transgenic carp fed diets with different protein levels. *Journal of Fish Biology*, 53, 115–129.

Fujita, H. and Yoshikawa, M. (1999). LKPNM: A prodrug-type ACE-inhibitory peptide derived from fish protein. *Immunopharmacology*, 44, 123–127.

Gao, J., Zhang, Y., Wang, H., Qin, S. (2005). Suspension culture of gametophytes of transgenic kelp in a photobioreactor. *Biotechnology Letters*, 27, 1025–1028.

Gao, J., Qin, S., Zhang, Y. (2006). Rapid optimization of process conditions for cultivation of transgenic *Laminaria japonica* gametophyte cells in a stirred-tank bioreactor. *Chemical Engineering Journal*, 122, 11–14.

GlobeNewswire (2021). AquaBounty Technologies announces results for the quarter and nine months ended September 30, 2021.

Hallmann, A. (2007). Algal transgenics and biotechnology. *Transgenic Plant Journal*, 1(1), 81–98.

Hew, C.L., Fletcher, G.L., Davies, P.L. (1995). Transgenic salmon: Tailoring the genome for food production. *Journal of Fish Biology*, 47, 1–19.

Higgs, D.A., Donaldson, E.M., Dye, H.M., McBride, J.R. (1976). Influence of bovine growth hormone and L-thyroxine on growth, muscle composition, and histological structure of the gonads, thyroid, pancreas, and pituitary of coho salmon (*Oncorhynchus kisutch*). *Journal of the Fisheries Board of Canada*, 33, 1585–1603.

Hobbie, K.R., DeAngelo, A.B., King, L.C., Winn, R.N., Law, J.M. (2009). Toward a molecular equivalent dose: Use of the medaka model in comparative risk assessment. *Comparative Biochemistry and Physiology Part C: Toxicology and Pharmacology*, 149, 141–151.

Horie, Y., Sugase, K., Horie, K. (1995). Physiological differences of soluble and insoluble dietary fibre fractions of brown algae and mushrooms in pepsin activity in vitro and protein digestibility. *Asia Pacific Journal of Clinical Nutrition*, 4(2), 251–255.

Houdebine, L.M. (2007). Que faire des saumons transgéniques à croissance accélérée ? *Cahiers Agricultures*, 16, 354–355.

Ifremer (2011). Aquaculture. Saumon atlantique [Online]. Available at: http://aquaculture.ifremer.fr/les-Filieres/Filiere-Poissons/La-decouverte-des-poissons/Saumon-atlantique.

Inoue, A., Kagaya, M., Ojima, T. (2008). Preparation of protoplasts from *Laminaria japonica* using native and recombinant abalone alginate lyases. *Journal of Applied Phycology*, 20, 633–640.

Inoue, A., Mashino, C., Kodama, T., Ojima, T. (2011). Protoplast preparation from *Laminaria japonica* with recombinant alginate lyase and cellulase. *Marine Biotechnology*, 13, 256–263.

Johnstone, R. (1985). Induction of triploidy in Atlantic salmon by heat shock. *Aquaculture*, 49, 133–139.

Joubert, Y. and Fleurence, J. (2008). Simultaneous extraction of proteins and DNA by an enzymatic treatment of the cell wall of *Palmaria palmata* (Rhodophyta). *Journal of Applied Phycology*, 20, 55–61.

Khubber, S., Marti-Quijal, F.J., Tomasevic, I., Remize, F., Barba, F.J. (2022). Lactic acid fermentation as a useful strategy to recover antimicrobial and antioxidant compounds from food and by-products. *Current Opinion in Food Science*, 43, 189–198.

Klemetsen, A., Amundsen, P.A., Dempson, J.B., Jonsson, B., Jonsson, N., O'Connell, M.F., Mortensen, E. (2003). Atlantic salmon *Salmo salar* L., brown trout *Salmo trutta* L. and Arctic charr *Salvelinus alpinus* (L.): A review of aspects of their life histories. *Ecology of Freshwater Fish*, 12(1), 1–59.

Kloareg, B. and Quatrano, R.S. (1988). Structure of the cell walls of marine algae and ecophysiological functions of the matrix polysaccharides. *Oceanography and Marine Biology: An Annual Review*, 26, 259–315.

Kobayashi, T., Kimura, B., Fujii, T. (2000). Strictly anaerobic halophiles isolated from canned Swedish fermented herrings (Surströmming). *International Journal of Food Microbiology*, 54(1/2), 81–89.

Kouakou, A.C., N'Guessan, F.K., Dadie, A.T., Montet, D., Djè, M.K. (2013). Production et commercialisation de l'adjuevan, poisson fermenté de Côte d'Ivoire. *Cahiers Agricultures*, 22, 559–567.

Lahaye, M. and Vigouroux, J. (1992). Liquefaction of dulse (*Palmaria palmata* (L.) Kuntze) by a commercial enzyme preparation and a purified endo, β-1, 4-D-xylanase. *Journal of Applied Phycology*, 4, 329–337.

Lestage, J.A. (1937). Le cannibalisme chez les truites : Réalité ou légende ? *Bulletin Français de Pisciculture*, 104, 190–193.

Lin, S. (2000). Transgenic zebrafish. In *Developmental Biology Protocols: Volume II*, Tuan, R.S. and Lo, C.W. (eds). Springer Publisher, 375–383.

Lin, C.M., Yarish, C., Chen, T. (2004). Development of a novel gene transfer method in *Porphyra*. *Bulletin of Fisheries Research Agency*, Supplement 1, 155.

Mabeau, S. and Fleurence, J. (1993). Seaweed in food products: Biochemical and nutritional aspects. *Trends in Food Science and Technology*, 4(4), 103–107.

Maclean, N. and Laight, R.J. (2000). Transgenic fish: An evaluation of benefits and risks. *Fish and Fisheries*, 1(2), 146–172.

Markert, J.R., Higgs, D.A., Dye, H.M., MacQuarrie, D.W. (1977). Influence of bovine growth hormone on growth rate, appetite, and food conversion of yearling coho salmon (*Oncorhynchus kisutch*) fed two diets of different composition. *Canadian Journal of Zoology*, 55, 74–83.

Marrion, O., Schwertz, A., Fleurence, J., Guéant, J.L., Villaume, C. (2003). Improvement of the digestibility of the proteins of the red alga *Palmaria palmata* by physical processes and fermentation. *Food/Nahrung*, 47, 339–344.

Martí-Quijal, F.J., Remize, F., Meca, G., Ferrer, E., Ruiz, M.J., Barba, F.J. (2020a). Fermentation in fish and by-products processing: An overview of current research and future prospects. *Current Opinion in Food Science*, 31, 9–16.

Martí-Quijal, F.J., Príncep, A., Tornos, A., Luz, C., Meca, G., Tedeschi, P., Ruiz, M.J., Barba, F.J., Mañes, J. (2020b). Isolation, identification and investigation of fermentative bacteria from sea bass (*Dicentrarchus labrax*): Evaluation of antifungal activity of fermented fish meat and by-products broths. *Foods*, 9, 576.

Miwa, S. and Inui, Y. (1986). Inhibitory effects of 17α-methyltestosterone and estradiol-17β on smoltification of sterilized amago salmon (*Oncorhynchus rhodurus*). *Aquaculture*, 53(1), 21–39.

Monteiro, P., Lomartire, S., Cotas, J., Pacheco, D., Marques, J.C., Pereira, L., Gonçalves, A.M. (2021). Seaweeds as a fermentation substrate: A challenge for the food processing industry. *Processes*, 9(11), 1953.

Muir, W.M. and Howard, R.D. (1999). Possible ecological risks of transgenic organism release when transgenes affect mating success: Sexual selection and the Trojan gene hypothesis. *Proceedings of the National Academy of Sciences*, 96, 13853–13856.

Nilsang, S., Lertsiri, S., Suphantharika, M., Assavanig, A. (2005). Optimization of enzymatic hydrolysis of fish soluble concentrate by commercial proteases. *Journal of Food Engineering*, 70, 571–578.

Nguyen, H.P.T., Morançais, M., Fleurence, J., Dumay, J. (2017). *Mastocarpus stellatus* as a source of R-phycoerythrin: Optimization of enzyme assisted extraction using response surface methodology. *Journal of Applied Phycology*, 29(3), 1563–1570.

Pérez, R. (1997). *Ces algues qui nous entourent: Conception actuelle, rôle dans la biosphère, utilisations, culture*. Quae, Versailles.

Qin, S., Jiang, P., Tseng, C. (2005). Transforming kelp into a marine bioreactor. *Trends in Biotechnology*, 23(5), 264–268.

Rajamuddin, M.A., Widyastut, U., Harris, E. (2016). Transformation of Kappa (κ)-Carrageenase gene from *Pseudoalteromonas* in seaweed *Kappaphycus alvarezii* [DOTY]. *Pakistan Journal of Biotechnology*, 13, 157–163.

Reboleira, J., Silva, S., Chatzifragkou, A., Niranjan, K., Lemos, M.F. (2021). Seaweed fermentation within the fields of food and natural products. *Trends in Food Science and Technology*, 116, 1056–1073.

Renshaw, S.A., Loynes, C.A., Trushell, D.M., Elworthy, S., Ingham, P.W., Whyte, M.K. (2006). A transgenic zebrafish model of neutrophilic inflammation. *Blood*, 108, 3976–3978.

Ryan, J.T., Ross, R.P., Bolton, D., Fitzgerald, G.F., Stanton, C. (2011). Bioactive peptides from muscle sources: Meat and fish. *Nutrients*, 3, 765–791.

Sager, J.J., Bai, Q., Burton, E.A. (2010). Transgenic zebrafish models of neurodegenerative diseases. *Brain Structure and Function*, 214(2), 285–302.

Shahidi, F. and Kamil, Y.J. (2001). Enzymes from fish and aquatic invertebrates and their application in the food industry. *Trends in Food Science and Technology*, 12, 435–464.

Sharma, N.K. and Rai, A.K. (2011). Algal particles in the Atmosphere. In *Encyclopedia of Environmental Health*, Nriagu, J.O. (ed.). Elsevier, London, 64–70.

Shelbourn, J.E., Clarke, W.C., McBride, J.R., Fagerlund, U.H., Donaldson, E.M. (1992). The use of 17α-methyltestosterone and 3, 5, 3′-triiodo-L-thyronine for sterilizing and accelerating the growth of zero-age coho salmon smolts (*Oncorhynchus kisutch*). *Aquaculture*, 103, 85–99.

Skåra, T., Axelsson, L., Stefánsson, G., Ekstrand, B., Hagen, H. (2015). Fermented and ripened fish products in the northern European countries. *Journal of Ethnic Foods*, 2, 18–24.

Stévant, P., Ólafsdóttir, A., Déléris, P., Dumay, J., Fleurence, J., Ingadóttir, B., Jonsdóttir, R., Ragueneau, E., Rebours, C., Rustad, T. (2020). Semi-dry storage as a maturation process for improving the sensory characteristics of the edible red seaweed dulse (*Palmaria palmata*). *Algal Research*, 51, 102048.

Sulistiani, E., Suharsono, S., Supena, E.D.J., Miftahudin, M. (2019). *Agrobacterium*-mediated genetic transformation of seaweed *Kappaphycus alvarezii* using Gα gene and callus cultures. *Biotropia*, 26, 40–50.

Tacon, A.J. (2003). Analyse des tendances de production en aquaculture. État de l'aquaculture dans le monde. Circulaire sur les pêches n° 886, Rév. 2. FAO.

Tanaka, M. and Kinoshita, M. (2001). Recent progress in the generation of transgenic medaka (*Oryzias latipes*). *Zoological Science*, 18(5), 615–622.

Toyohara, H., Nakata, T., Touhata, K., Hashimoto, H., Kinoshita, M., Sakaguchi, M., Nishikimi, M., Yagi, K., Wakatmatsu, Y, Ozato, K. (1996). Transgenic expression of L-gulono-γ-lactone Oxidase in médaka (*Oryzias latipes*), a teleost fish that lacks this enzyme necessary for L-ascorbic acid biosynthesis. *Biochemical and Biophysical Research Communications*, 223, 650–653.

Tsuda, H., Kubota, K., Matsumoto, T., Ishimi, Y. (2012). Isolation and identification of lactic acid bacteria in traditional fermented sushi, funazushi, from Japan. *Food Science and Technology Research*, 18(1), 77–82.

Uchida, M. and Miyoshi, T. (2013). Algal fermentation – The seed for a new fermentation industry of foods and related products. *Japan Agricultural Research Quarterly*, 47(1), 53–63.

Vera, J., Castro, J., Gonzalez, A., Moenne, A. (2011). Seaweed polysaccharides and derived oligosaccharides stimulate defense responses and protection against pathogens in plants. *Marine Drugs*, 9, 2514–2525.

Winn, R.N., Norris, M.B., Brayer, K.J., Torres, C., Muller, S.L. (2000). Detection of mutations in transgenic fish carrying a bacteriophage λ cII transgene target. *Proceedings of the National Academy of Sciences*, 97, 12655–12660.

Winn, R.N., Majeske, A.J., Jagoe, C.H., Glenn, T.C., Smith, M.H., Norris, M.B. (2008). Transgenic λ medaka as a new model for germ cell mutagenesis. *Environmental and Molecular Mutagenesis*, 49, 173–184.

Zhang, P., Hayat, M., Joyce, C., Gonzalez-Villaseñor, L.I., Lin, C.M., Dunham, R. A., Chen, T.T., Powers, D.A. (1990). Gene transfer, expression and inheritance of pRSV-rainbow trout-GH cDNA in the common carp, *Cyprinus carpio* (Linnaeus). *Molecular Reproduction and Development*, 25, 3–13.

Zhao, Y., Wang, C., Xia, S., Jiang, J., Hu, R., Yuan, G., Hu, J. (2014). Biosensor medaka for monitoring intersex caused by estrogenic chemicals. *Environmental Science and Technology*, 48, 2413–2420.

Index

Other titles from

iSTE

in

Ecological Science

2023

GAID Kader
Drinking Water Treatment 1: Water Quality and Clarification
Drinking Water Treatment 2: Chemical and Physical Elimination of Organic Substances and Particles
Drinking Water Treatment 3: Organic and Mineral Micropolluants
Drinking Water Treatment 4: Membranes Applied to Drinking Water and Desalination
Drinking Water Treatment 5: Calco-carbonic Equilibrium and Disinfection

2022

AMIARD Jean-Claude
Management of Radioactive Waste
(Radioactive Risk Set – Volume 5)
Marine Radioecology
(Radioactive Risk Set – Volume 6)

GODET Laurent, DUFOUR Simon, ROLLET Anne-Julia
The Baseline Concept in Biodiversity Conservation: Being Nostalgic or Not in the Anthropocene Era

GRISON Claude, CASES Lucie, LE MOIGNE Mailys, HOSSAERT-MCKEY Martine
Photovoltaism, Agriculture and Ecology From Agrivoltaism to Ecovoltaism

LÉVÊQUE Christian
Biodiversity Erosion: Issues and Questions

2021

FLEURENCE Joël
Microalgae: From Future Food to Cellular Factory

AMIARD Jean-Claude
Disarmament and Decommissioning in the Nuclear Domain (Radioactive Risk Set – Volume 4)

JOLY Fernand, BOURRIÉ Guilhem
Mankind and Deserts 1: Deserts, Aridity, Exploration and Conquests
Mankind and Deserts 2: Water and Salts
Mankind and Deserts 3: Wind in Deserts and Civilizations

MUTTIN Frédéric, THOMAS Hélène
Marine Environmental Quality

PIOCH Sylvain, SOUCHE Jean-Claude
Eco-design of Marine Infrastructures: Towards Ecologically-informed Coastal and Ocean Development

2020

BRUSLÉ Jacques, QUIGNARD Jean-Pierre
Fish Behavior 1: Eco-ethology
Fish Behavior 2: Ethophysiology

LE FLOCH Stéphane
Remote Detection and Maritime Pollution: Chemical Spill Studies

MIGON Christophe, NIVAL Paul, SCIANDRA Antoine
The Mediterranean Sea in the Era of Global Change 1: 30 Years of Multidisciplinary Study of the Ligurian Sea
The Mediterranean Sea in the Era of Global Change 2: 30 Years of Multidisciplinary Study of the Ligurian Sea

ROSSIGNOL Jean-Yves
Climatic Impact of Activities: Methodological Guide for Analysis and Action

2019

AMIARD Jean-Claude
Industrial and Medical Nuclear Accidents: Environmental, Ecological, Health and Socio-economic Consequences (Radioactive Risk Set – Volume 2)
Nuclear Accidents: Prevention and Management of an Accidental Crisis (Radioactive Risk Set – Volume 3)

BOULEAU Gabrielle
Politicization of Ecological Issues: From Environmental Forms to Environmental Motives

DAVID Valérie
Statistics in Environmental Sciences

GIRAULT Yves
UNESCO Global Geoparks: Tension Between Territorial Development and Heritage Enhancement

KARA Mohamed Hichem, QUIGNARD Jean-Pierre
Fishes in Lagoons and Estuaries in the Mediterranean 2: Sedentary Fish
Fishes in Lagoons and Estuaries in the Mediterranean 3A: Migratory Fish
Fishes in Lagoons and Estuaries in the Mediterranean 3B: Migratory Fish

OUVRARD Benjamin, STENGER Anne
Incentives and Environmental Policies: From Theory to Empirical Novelties

2018

AMIARD Jean-Claude
Military Nuclear Accidents: Environmental, Ecological, Health and Socio-economic Consequences
(Radioactive Risk Set – Volume 1)

FLIPO Fabrice
The Coming Authoritarian Ecology

GUILLOUX Bleuenn
Marine Genetic Resources, R&D and the Law 1: Complex Objects of Use

KARA Mohamed Hichem, QUIGNARD Jean-Pierre
Fishes in Lagoons and Estuaries in the Mediterranean 1: Diversity, Bioecology and Exploitation

2016

BAGNÈRES Anne-Geneviève, HOSSAERT-MCKEY Martine
Chemical Ecology

2014

DE LARMINAT Philippe
Climate Change: Identification and Projections

Printed and bound by CPI Group (UK) Ltd, Croydon, CR0 4YY
13/09/2026
14961049-0001